Collins

2026 GUIDE
to the
NIGHT SKY

Radmila Topalovic and Dominic Ford

Published by Collins
An imprint of HarperCollins Publishers
1 Robroyston Gate,
Glasgow G33 1JN
collins.reference@harpercollins.co.uk
collins.co.uk

HarperCollins Publishers
Macken House
39/40 Mayor Street Upper
Dublin 1
D01 C9W8
Ireland

In association with
Royal Museums Greenwich, the group name for the National Maritime Museum,
Royal Observatory Greenwich, Queen's House and *Cutty Sark*
www.rmg.co.uk

First published 2025

© HarperCollins Publishers 2025
Text © Radmila Topalovic 2025
Diagrams © Dominic Ford 2025
Photographs © see credits page 110

Collins® is a registered trademark of HarperCollins Publishers Ltd

All rights reserved. No part of this publication may be reproduced, stored in a retrieval system, or transmitted, in any form or by any means, electronic, mechanical, photocopying, recording or otherwise without the prior permission in writing of the publisher and copyright owners.

Without limiting the exclusive rights of any author, contributor or the publisher of this publication, any unauthorised use of this publication to train generative artificial intelligence (AI) technologies is expressly prohibited. HarperCollins also exercise their rights under Article 4(3) of the Digital Single Market Directive 2019/790 and expressly reserve this publication from the text and data mining exception.

HarperCollins does not warrant that any website mentioned in this title will be provided uninterrupted, that any website will be error free, that defects will be corrected, or that the website or the server that makes it available are free of viruses or bugs. For full terms and conditions please refer to the site terms provided on the website.

The publishers wish to acknowledge their deep debt of gratitude to two of their late astronomy authors: Storm Dunlop FRAS, FRMetS (1942-2025) for his stewardship of the Guides to the Night Sky and Night Sky Almanac and the enormous contribution of his friend and co-author the late Wil Tirion (1943-2024) for his expert star charts and diagrams. They both leave a valuable and ongoing legacy.

A catalogue record for this book is available from the British Library

ISBN 978 0 00 874766 4

10 9 8 7 6 5 4 3 2 1

Printed in Italy by Rotolito

If you would like to comment on any aspect of this book, please contact us at the above address or online.
e-mail: collins.reference@harpercollins.co.uk

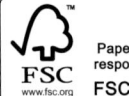

MIX
Paper | Supporting
responsible forestry
FSC
www.fsc.org   FSC™ C007454

This book contains FSC™ certified paper and other controlled sources to ensure responsible forest management.

For more information visit: www.harpercollins.co.uk/green

Contents

4   Introduction

The Constellations
10   The Northern Circumpolar Constellations
12   The Winter Constellations
13   The Spring Constellations
14   The Summer Constellations
15   The Autumn Constellations

The Moon and the Planets
16   The Moon
18   Map of the Moon
20   Eclipses
22   The Planets
26   Minor Planets
28   Comets

30   Introduction to the Month-by-Month Guide

Month-by-Month Guide
34   January
40   February
46   March
52   April
58   May
64   June
70   July
76   August
82   September
88   October
94   November
100   December

106   Dark Sky Sites
108   Glossary and Tables
110   Acknowledgements
111   Further Information

Introduction

The aim of this Guide is to help people find their way around the night sky at any time of the year, by showing how the stars that are visible change from month to month, and by highlighting various events that occur during 2026. The objects and events described may be observed with the naked eye, or nothing more complicated than a pair of binoculars.

The conditions for observing naturally vary over the course of the year. During the summer, twilight may persist throughout the night and make it difficult to see the faintest stars. There are three recognized stages of twilight: civil twilight, when the Sun is less than 6° below the horizon; nautical twilight, when the Sun is between 6° and 12° below the horizon; and astronomical twilight, when the Sun is between 12° and 18° below the horizon. Full darkness occurs only when the Sun is more than 18° below the horizon. During nautical twilight, only the very brightest stars are visible. During astronomical twilight, the faintest stars visible to the naked eye may be seen directly overhead, but are lost at lower altitudes. As the diagram shows, full darkness never occurs during June and most of July at the latitude of London, and at Edinburgh nautical twilight persists throughout the whole night.

Moonlight will affect the visibility of objects, providing a natural form of light pollution. At Full Moon, it may be very difficult to see some of the fainter stars and objects, and even when the Moon is at a smaller phase it may seriously interfere with visibility if it is near the stars or planets in which you are interested. A full lunar calendar is given for each month and may be used to plan night sky observations. It may also be useful to look at moonrise and moonset times for your particular location.

The celestial sphere

All the objects in the sky (including the Sun, Moon and stars) appear to lie at some indeterminate distance on a large sphere, centred on the Earth. This *celestial sphere* has various reference points and features that are related to those of the Earth. If the Earth's rotational axis (an imaginary line through the North and South Poles and the core of the Earth) is extended, for example, it points to the North and South Celestial Poles. Similarly, the *celestial equator* lies in the same plane as the Earth's equator, it divides the sky into northern and southern hemispheres. This Guide is written for use in Britain and Ireland, therefore

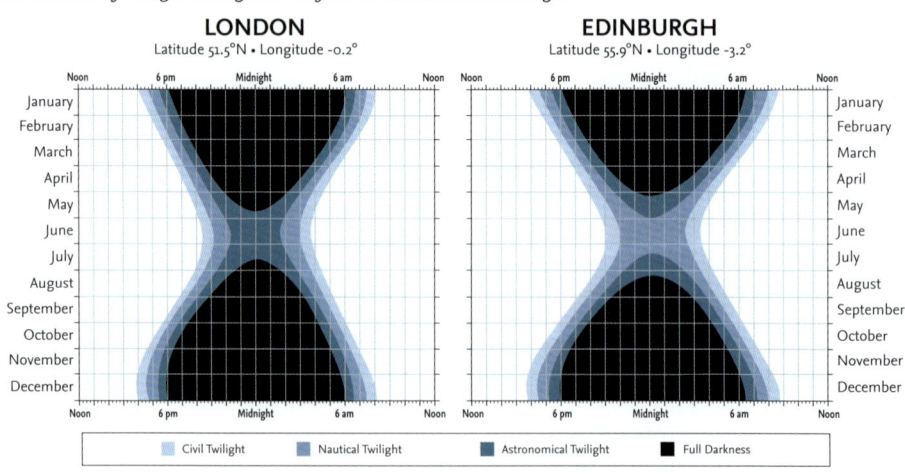

The duration of twilight throughout the year at London and Edinburgh.

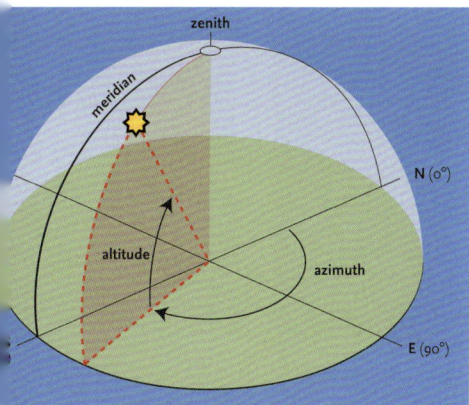
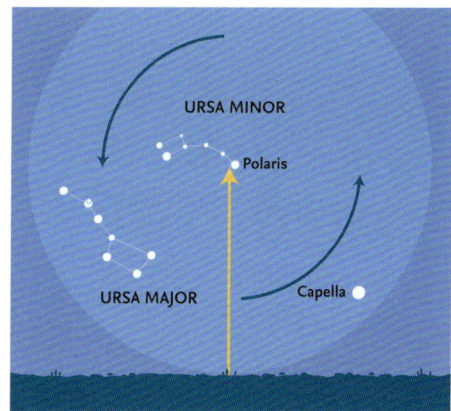

Measuring altitude and azimuth on the celestial sphere.   *The altitude of the North Celestial Pole equals the observer's latitude.*

the area of the sky that it describes includes the whole of the northern celestial hemisphere and those portions of the southern sky that become visible at different times of the year. Stars in the far south, however, remain below the horizon throughout the year, and are not included.

It is useful to know some of the astronomy terms for various parts of the sky. As seen by an observer, half of the celestial sphere is invisible at any point in time; these objects will be below the horizon. The point directly overhead is known as the **zenith**, and the (invisible) one below one's feet as the **nadir**. The line running from the north point on the horizon, up through the zenith and then down to the south point is the **meridian**. This is an important invisible line in the sky, because objects are highest in the sky, and thus easiest to see, when they cross the meridian in the south. Objects are said to **transit** when they cross this line in the sky.

In this book, reference is frequently made in the text and in the diagrams to the standard compass points around the horizon. The position of any object in the sky specific to the observer's location on Earth may be described by its **altitude** (measured in degrees above the horizon), and its **azimuth** (measured in degrees from north 0°, through east 90°, south 180° and west 270°). Experienced amateurs and professional astronomers also use another system of specifying locations on the celestial sphere called right ascension and declination, but here the simpler method will suffice.

The celestial sphere appears to rotate about an invisible axis, running between the North and South Celestial Poles. The location (i.e. the altitude) of the Celestial Poles depends entirely on the observer's latitude on Earth. At the North Pole (latitude 90°), the North Celestial Pole (NCP) would be directly overhead (at the zenith, or at an altitude of 90°). The charts in this book are produced for the latitude of 50°N, so the NCP is 50° above the northern horizon. The fact that the NCP is fixed relative to the horizon means that all the stars within 50° of the pole are always above the horizon and may, therefore, always be seen at night, regardless of the time of year. This northern circumpolar region is an ideal place to begin learning the sky, and ways to identify the circumpolar stars and constellations will be described shortly.

The ecliptic and the zodiac

Another important line on the celestial sphere is the Sun's apparent path against the background stars – in reality the result of the Earth's orbit around the Sun. This is known as the *ecliptic*. The point where the Sun, apparently moving along the ecliptic, crosses the celestial equator from south to north is known as the

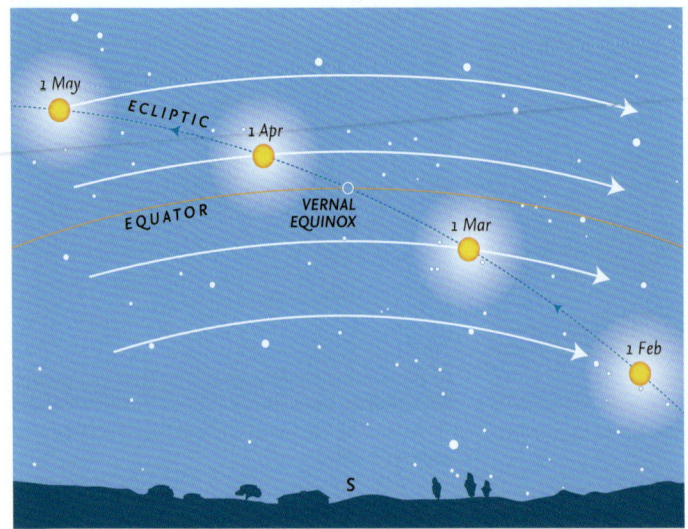

The Sun crossing the celestial equator in spring.

vernal (or spring) equinox, which occurs on 20 March. At this time (and at the autumnal equinox, on 22 or 23 September, when the Sun crosses the celestial equator from north to south) day and night are almost exactly equal in length. The vernal equinox is currently located in the constellation of Pisces; this means the Sun is moving through Pisces on this day. It is important in astronomy because it defines the zero point for the system of celestial coordinates (right ascension) not used in this Guide.

The Moon and planets are to be found in a band of sky all the way around the celestial sphere that extends 8° on either side of the ecliptic. This is because the orbits of the Moon and planets are inclined at various angles to the ecliptic (i.e. to the plane of the Earth's orbit). This band of sky is known as the zodiac and, when originally devised, consisted of twelve **constellations**, all of which were considered to be exactly 30° wide. When the constellation boundaries were formally established by the International Astronomical Union in 1930, adjustments were made resulting in the ecliptic passing through thirteen constellations and the Moon and planets passing through several other constellations that are adjacent to the original twelve.

The constellations

Since ancient times, the celestial sphere has been divided into various constellations, most dating back to antiquity and usually associated with certain myths or legendary people and animals. The boundaries of the 88 constellations across the whole celestial sphere have been fixed by international agreement and their names (in Latin) are largely derived from Greek or Roman originals. Some of the names of the most prominent stars are of Greek or Roman origin, but many are derived from Arabic names. Many bright stars have no individual names and, for many years, stars were identified by terms such as 'the star in Hercules' right foot'. A more sensible scheme was introduced by the German astronomer Johannes Bayer in the early seventeenth century. Following his scheme – which is still used today – most of the brightest stars are identified by a Greek letter followed by the genitive form of the constellation's Latin name. An example is the Pole Star, also known as Polaris and α Ursae Minoris (abbreviated α UMi). The Greek alphabet is shown on page 109, along with a list of all the constellations that may be seen from latitude 50°N, together with abbreviations, their genitive forms and

English names. Other naming schemes exist for fainter stars; these are not used in this book.

Asterisms

Apart from the constellations, certain groups of stars, which may form a part of a larger constellation or cross several constellations, are readily recognizable and have been given individual names. These groups are known as *asterisms*, and the most famous (and well-known) is 'the Plough', the common name for the seven brightest stars in the constellation of *Ursa Major*, the Great Bear. The names and details of some asterisms mentioned in this book are given in the list on page 110.

Magnitudes

The brightness of a star, planet or other body is given in magnitudes (mag.). This is a mathematically defined scale where larger (positive) numbers indicate a fainter object. The scale extends beyond the zero point to negative numbers for very bright objects. (Sirius, the brightest star in the night sky is mag. -1.4.) Most observers are able to see stars as faint as about mag. 6, under very clear skies.

The Moon

Although the daily spin of the Earth carries the sky from east to west (objects in the south rise in the east and set in the west) the Moon gradually moves eastwards relative to the background stars by approximately its diameter in an hour. This is equivalent to about half a degree across the sky. Its apparent eastward motion and change in phase are caused by its orbit around the Earth and changing position relative to the Sun.

Normally, in its orbit around the Earth, the Moon passes above or below the direct line between the Earth and the Sun (at New Moon) or outside the area obscured by the Earth's shadow (at Full Moon), explained in the diagram on page 17. Occasionally, however, the three bodies are more or less perfectly aligned to give an *eclipse*: a solar eclipse at New Moon or a lunar eclipse at Full Moon. Depending on the exact circumstances, a solar eclipse may be merely partial (when the Moon does not cover the whole of the Sun's disk); annular (when the Moon is too far from Earth in its orbit to appear large enough to hide the whole of the Sun); or total. Total and annular eclipses are visible from very restricted areas of the Earth, but partial eclipses are normally visible over a wider area.

Somewhat similarly, at a lunar eclipse, the Moon may pass through the lighter outer zone of the Earth's shadow, the **penumbra**, called a penumbral eclipse, which is not generally perceptible to the naked eye; part of the Moon could pass within the darkest part of the Earth's shadow, the *umbra*, in a partial eclipse; or it could move completely within the umbra in a total eclipse. Unlike solar eclipses, lunar eclipses are visible from large areas of the Earth.

Occasionally, as it moves across the sky, the Moon passes between the Earth and individual planets or distant stars, giving rise to an *occultation*. As with solar eclipses, such occultations are visible from restricted areas of the world.

The planets

The planets are always moving against the background stars, therefore they are treated in some detail in the monthly pages and information is given regarding when they are close to the Sun, other planets, the Moon or any of five bright stars that lie near the ecliptic. Such events are known as *appulses* or, more frequently, as *conjunctions*. (There are technical differences in the way these terms are defined – and should be used – in astronomy, but these need not concern us here.) The positions of the planets are shown for every month on a special chart of the ecliptic.

The conditions of most favourable visibility depend on whether the planet is one of the two known as *inferior planets* – planets that orbit the Sun within the Earth's orbit (Mercury and Venus) – or one of the five *superior planets* – planets that orbit the Sun beyond the Earth. Of the latter, three (Mars, Jupiter and Saturn) are covered in detail, these are visible to the

INTRODUCTION   7

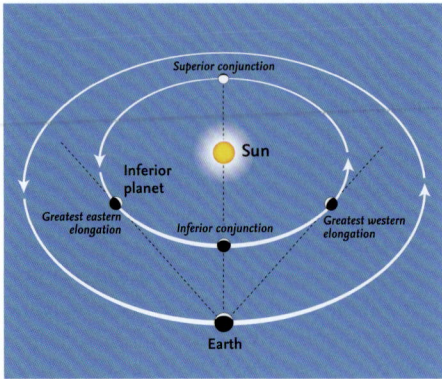

Inferior planet.

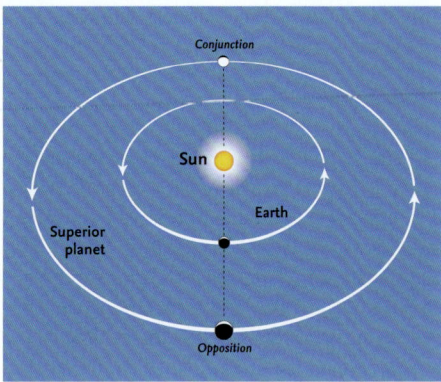

Superior planet.

naked eye. Occasionally, details of the fainter superior planets, Uranus and Neptune, are included, and special charts are given for them on page 25.

The inferior planets are most readily seen at eastern or western elongation, when their angular distance from the Sun (separation from the Sun in degrees) is greatest. Superior planets are best seen at **opposition**, when they are directly opposite the Sun in the sky and cross the meridian at local midnight.

Angular distance can be measured approximately by holding one hand at arm's length. The various angles are shown in the diagram, together with the separations of the various stars in the Plough.

Meteors

At some time or other, nearly everyone has seen a meteor – a 'shooting star' – as it flashed across the sky. The particles that cause meteors – known technically as 'meteoroids' – normally range in size from that of a grain of sand (or even smaller) to the size of a pea. **Fireballs** or **bolides** are very bright meteors (brighter than magnitude -4) that are caused by objects up to 1 metre in size. Fireballs sometimes cause sonic booms that may be heard some time after the meteor is seen. On any night of the year there are occasional meteors, known as sporadics, that may travel in any direction. These occur at a rate that is normally between three and eight per hour.

Far more important, however, are meteor showers, which occur at fixed periods of the year, when the Earth encounters a trail of particles left behind by a comet or, very occasionally, by a minor planet (asteroid). Meteors always appear to diverge from a single point on the sky, known as the radiant, and the radiants of major showers are shown on the charts. Meteors that come from a circular area 8° in diameter around the radiant are classed as belonging to the particular shower. All others that do not come from that area are sporadics (or, occasionally from another shower that is active at the same time). A list of the major meteor showers is given on page 31. Examples of meteors are shown on pages 32, 53, 77 and 101.

Looking directly at the radiant is not the most effective way of seeing meteors. They are most likely to be noticed if one is looking about 40–45° away from the radiant position. This is approximately two hand-spans as shown in the diagram for measuring angles.

Other objects

In the late eighteenth century Charles Messier, a French astronomer, compiled a catalogue of objects while he was searching for comets. These nebulae, clusters of stars and galaxies were given 'Messier numbers' (some already had names given several thousand years ago such as Praesepe – the Beehive Cluster). Some, such as the Andromeda Galaxy, M31, and the

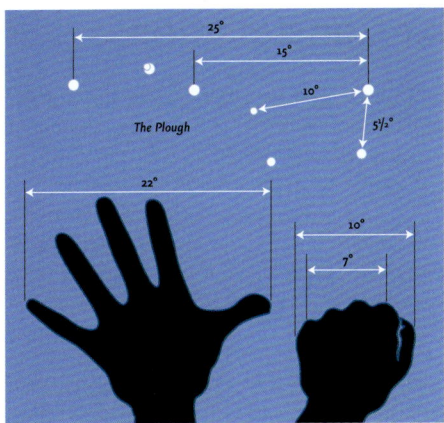
Measuring angles in the sky.

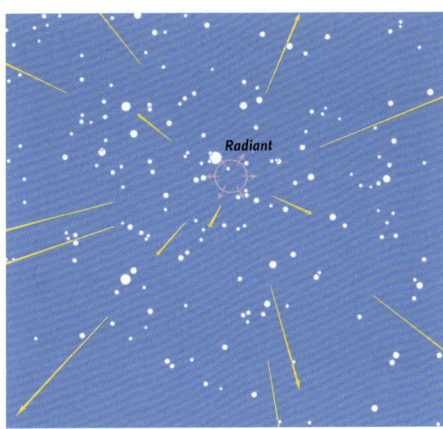
Meteor shower (showing the April Lyrid radiant).

Orion Nebula, M42, may be seen by the naked eye, but all those given in the list will benefit from the use of binoculars. Apart from galaxies, such as M31, which contain thousands of millions of stars, there are also two types of cluster: open clusters, such as M45, the Pleiades, which may consist of a few dozen to some hundreds of stars; and globular clusters, such as M13 in Hercules, which are spherical concentrations of many thousands of stars. One or two gaseous nebulae, consisting of gas illuminated by stars within them, are also visible. The Orion Nebula, M42, is one, and is illuminated by a group of four stars, known as the Trapezium, which may be seen by using a good pair of binoculars.

Some interesting objects.

Messier / NGC	Name	Type	Constellation	Maps (months)
—	Hyades	open cluster	Taurus	Sep–Mar
—	Double Cluster	open cluster	Perseus	All year
—	Melotte 111 *(Coma Cluster)*	open cluster	Coma Berenices	Jan–Aug
M3	—	globular cluster	Canes Venatici	Feb–Aug
M4	—	globular cluster	Scorpius	Jun–Jul
M8	Lagoon Nebula	gaseous nebula	Sagittarius	Jul–Aug
M11	Wild Duck Cluster	open cluster	Scutum	Jun–Oct
M13	Hercules Cluster	globular cluster	Hercules	Mar–Oct
M15	—	globular cluster	Pegasus	Jun–Dec
M20	Trifid Nebula	gaseous nebula	Sagittarius	Jun–Aug
M22	—	globular cluster	Sagittarius	Jun–Aug
M27	Dumbbell Nebula	planetary nebula	Vulpecula	May–Dec
M31	Andromeda Galaxy	galaxy	Andromeda	Jun–Mar
M35	—	open cluster	Gemini	Oct–Apr
M42	Orion Nebula	gaseous nebula	Orion	Nov–Mar
M44	Praesepe	open cluster	Cancer	Nov–Jun
M45	Pleiades	open cluster	Taurus	Sep–Mar
M57	Ring Nebula	planetary nebula	Lyra	Apr–Nov
M67	—	open cluster	Cancer	Dec–May
NGC 752	—	open cluster	Andromeda	Jul–Mar
NGC 3242	Ghost of Jupiter	planetary nebula	Hydra	Feb–May

INTRODUCTION

The Northern Circumpolar Constellations

The northern circumpolar stars are the key to starting to identify the constellations. For anyone in the northern hemisphere they are visible at any time of the year, and nearly everyone is familiar with the seven stars of the Plough – known as the Big Dipper in North America – an asterism that forms part of the large constellation of **Ursa Major** (the Great Bear).

Ursa Major

Due to the Earth's orbit around the Sun, Ursa Major lies in different parts of the evening sky at different periods of the year. The diagram below shows its position at the beginning of the four main seasons. The seven stars of the Plough remain visible throughout the year anywhere north of latitude 40°N. Even at the latitude (50°N) for which the charts in this book are drawn, many of the stars in the southern portion of the constellation of Ursa Major are hidden below the horizon for part of the year or (particularly in late summer) are too faint to be seen late in the night.

Polaris and Ursa Minor

The two stars **Dubhe** and **Merak** (α and β Ursae Majoris, respectively), farthest from the 'tail' are known as 'the Pointers'. A line from Merak to Dubhe, extended about five times their separation, leads to the Pole Star, **Polaris**, or α Ursae Minoris. All the stars in the northern sky appear to rotate around it. There are five main stars in the constellation of **Ursa Minor**, and the two farthest from the Pole, **Kochab** and **Pherkad** (β and γ Ursae Minoris, respectively), are known as 'the Guards'.

Cassiopeia

Facing Ursa Major on the other side of Polaris lies **Cassiopeia**. It is highly distinctive, appearing as five stars forming a letter 'W' or 'M' depending on its orientation. Provided the sky is reasonably clear of clouds, you will nearly always be able to see either Ursa Major or Cassiopeia, and thus be able to orientate yourself on the sky.

To find Cassiopeia, start with **Alioth** (ε Ursae Majoris), the first star in the tail of the Great Bear. A line from this star extended through Polaris points directly towards γ Cassiopeiae, the central star of the five.

Cepheus

Although the constellation of **Cepheus** is fully circumpolar, it is not nearly as well-known as Ursa Major, Ursa Minor or Cassiopeia, partly because its stars are fainter. Its shape is rather like the gable end of a house. The line from the Pointers through Polaris, if extended, leads to **Errai** (γ Cephei) at the 'top' of the 'gable'. The brightest star, **Alderamin** (α Cephei) lies in the Milky Way region, at the 'bottom right-hand corner' of the figure.

Draco

The constellation of **Draco** consists of a quadrilateral of stars, known as 'the Head of Draco' (and also 'the Lozenge'), and a long chain of stars forming the neck and body of the dragon. To find the Head of Draco, locate the two stars **Phecda** and **Megrez** (γ and δ Ursae Majoris) in the Plough, opposite the Pointers.

Extend a line from Phecda through Megrez by about eight times their separation, right

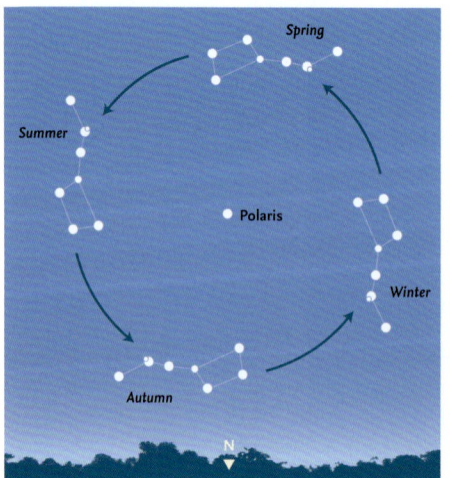

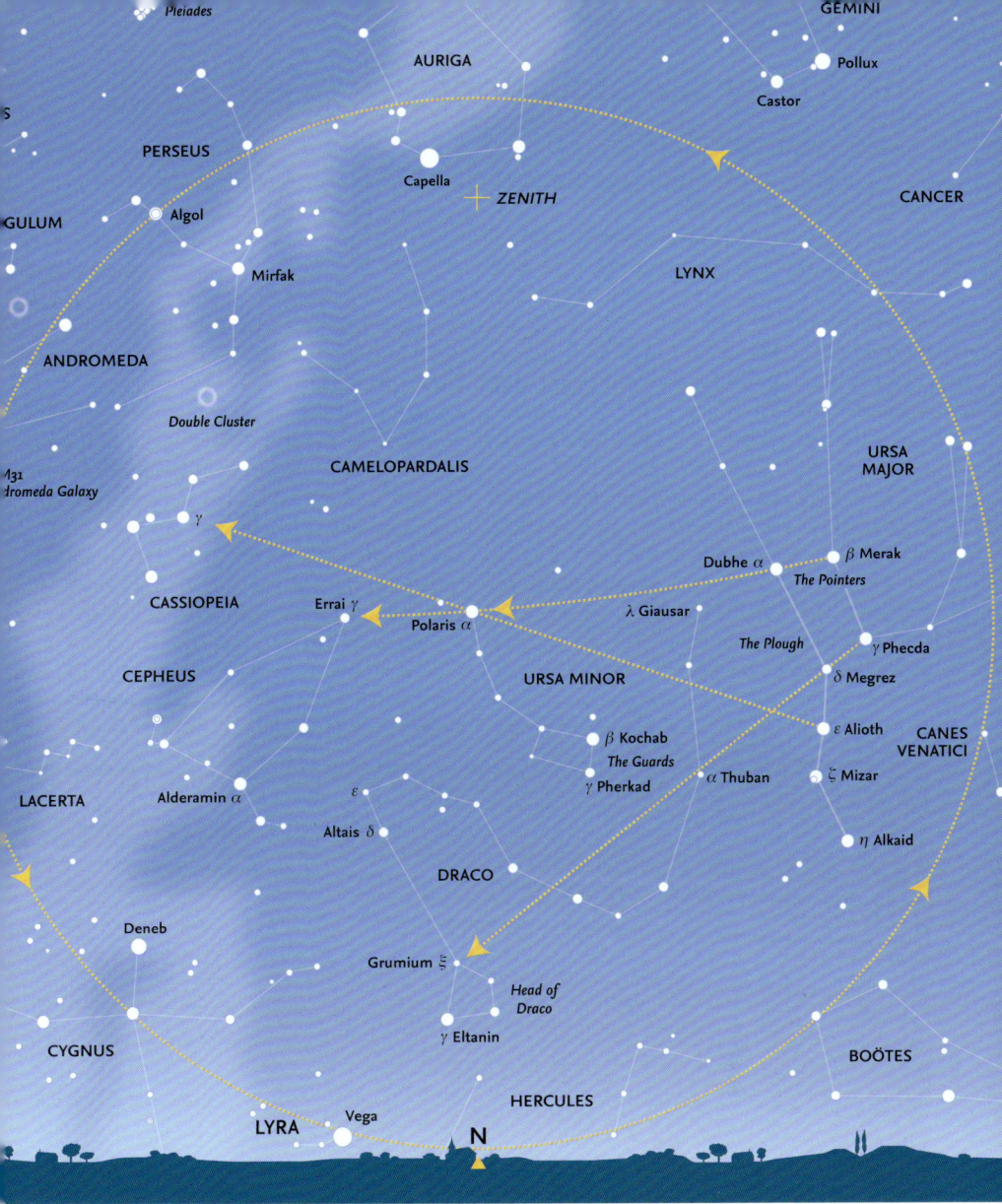

The stars and constellations inside the circle are always above the horizon, seen from our latitude.

across the sky below the Guards in Ursa Minor, ending at **Grumium** (ξ Draconis) at one corner of the quadrilateral. The brightest star, **Eltanin** (γ Draconis), lies farther to the south. From the head of Draco, the constellation first runs northwest to **Altais** (δ Draconis) and ε Draconis, then doubles back southwards before winding its way through **Thuban** (α Draconis) and ending at **Giausar** (λ Draconis) almost between the Pointers and Polaris.

THE CONSTELLATIONS 11

The Winter Constellations

The winter sky is dominated by several bright stars and distinctive constellations. The most conspicuous constellation is **Orion** (the Hunter), the main body of which has an hourglass shape. It straddles the celestial equator and is thus visible from anywhere in the world. The three stars that form 'the Belt' of Orion point down towards the southeast and to **Sirius** (α Canis Majoris), the brightest star in the night sky. **Mintaka** (δ Orionis), the star at the northeastern end of the Belt, farthest from Sirius, actually lies just slightly south of the celestial equator.

The bright red supergiant star **Betelgeuse** (α Orionis) lies above the belt, as does **Bellatrix** (γ Orionis). A line from Bellatrix at the 'top right-hand corner' of Orion, through **Aldebaran** (α Tauri), past the 'V' of the Hyades cluster, points to the distinctive cluster of bright blue stars known as the **Pleiades**, or 'the Seven Sisters'. Aldebaran is one of the five bright stars that may sometimes be occulted (hidden) by the Moon. On the far eastern side of Orion sits the constellation of **Leo** (the lion), a prominent constellation in the spring sky, easy to find with its backwards question mark (the 'head' of the lion).

Six bright stars in six different constellations: **Capella** (α Aurigae), **Aldebaran** (α Tauri), **Rigel** (β Orionis), Sirius (α Canis Majoris), **Procyon** (α Canis Minoris) and **Pollux** (β Geminorum) form what is sometimes known as 'the Winter Hexagon'. Pollux is accompanied to the northwest by the slightly fainter star of **Castor** (α Geminorum), the second 'Twin' of Gemini.

In a counterpart to the famous asterism 'the Summer Triangle', an almost perfect equilateral triangle, 'the Winter Triangle', is formed by Betelgeuse (α Orionis), Sirius (α Canis Majoris) and Procyon (α Canis Minoris).

Several of the stars in this region of the sky show distinctive tints: Betelgeuse (α Orionis) is reddish, Aldebaran (α Tauri) is orange and Rigel (β Orionis) is blue-white.

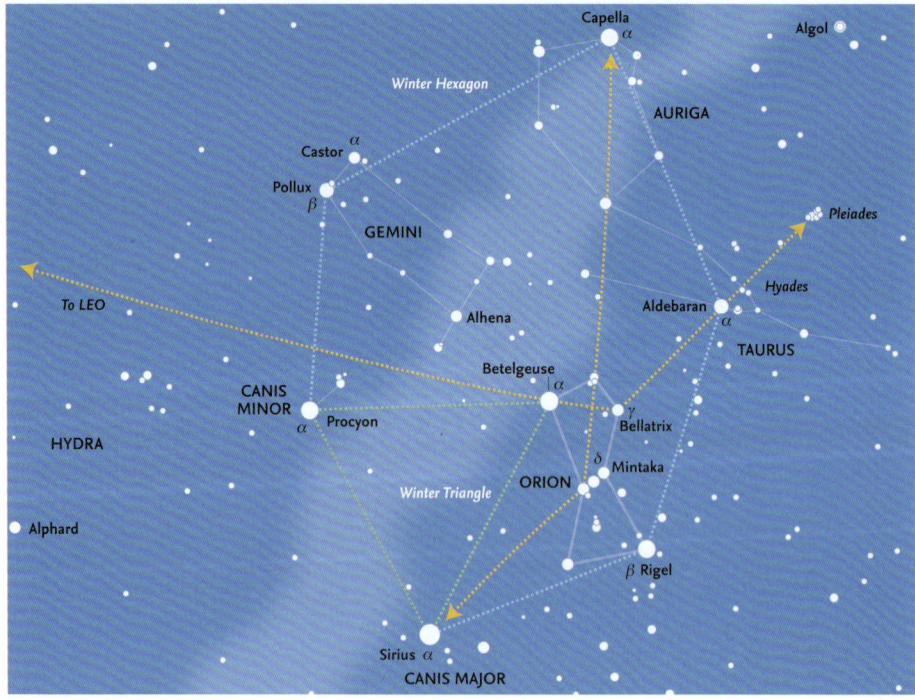

The Spring Constellations

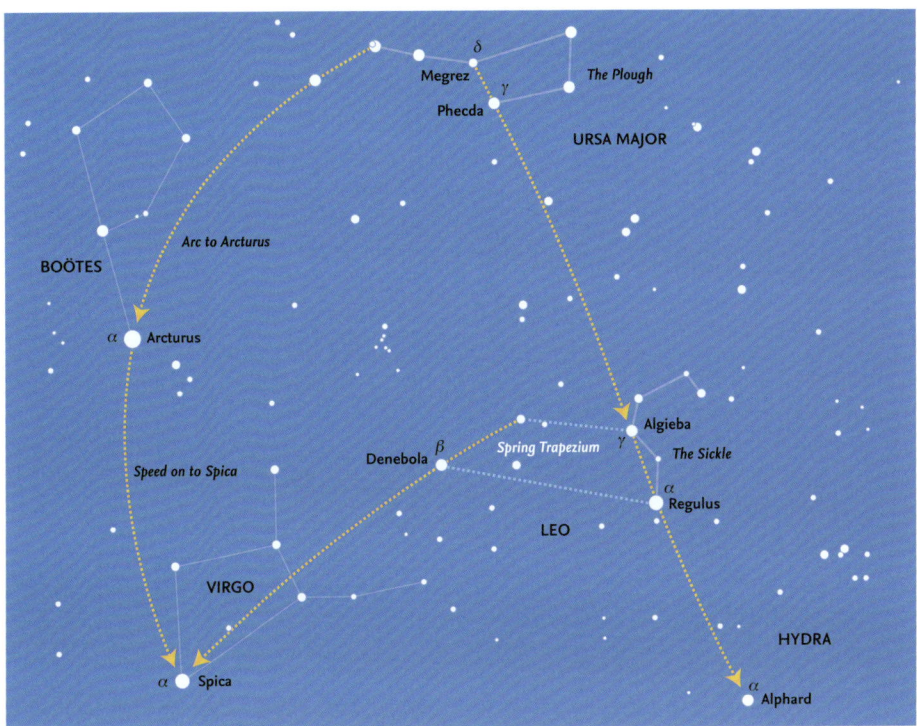

The most prominent constellation in the spring sky is the zodiacal constellation of **Leo**, and its brightest star, **Regulus** (α Leonis), which may be found by extending a line from Megrez and Phecda (δ and γ Ursae Majoris, respectively) – the two stars on the opposite side of the bowl of the Plough from the Pointers – down to the southeast. Regulus forms the 'dot' of the backwards question mark known as 'the Sickle'. Regulus, like Aldebaran in Taurus, is one of the bright stars that lie close to the ecliptic, and that are occasionally occulted by the Moon. The same line from Ursa Major to Regulus, if continued, leads to **Alphard** (α Hydrae), the brightest star in **Hydra**, the largest of the 88 constellations.

The shape formed by the body of Leo is sometimes known as 'the Spring Trapezium'. At the other end of the constellation from Regulus is **Denebola** (β Leonis), and the line forming the back of the constellation through Denebola points to the bright star **Spica** (α Virginis) in the constellation of **Virgo**, a rough quadrilateral of moderately bright stars and fainter lines of stars extending outwards. A saying that helps to locate Spica is well-known to astronomers: 'Arc to Arcturus and then speed on to Spica.' This suggests following the arc of the tail of Ursa Major to Arcturus and then a straight line on to Spica. **Arcturus** (α Boötis) is actually the brightest star in the northern hemisphere of the sky. (Although other stars, such as Sirius, are brighter, they are all in the southern hemisphere.) Overall, the constellation of **Boötes** is sometimes described as 'kite-shaped' or 'shaped like the letter P'.

THE CONSTELLATIONS 13

The Summer Constellations

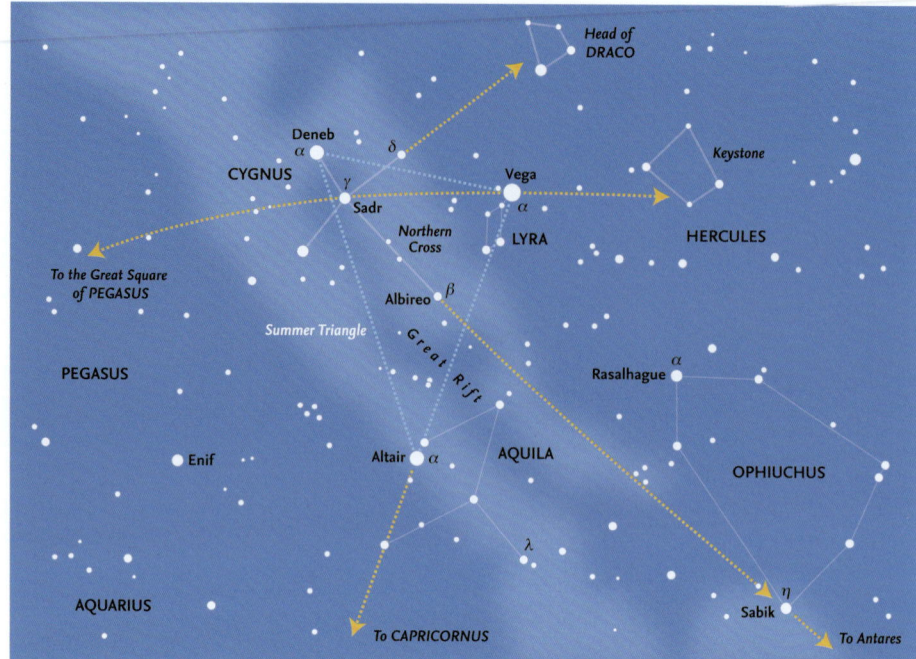

On summer nights, the three bright stars **Deneb** (α Cygni), **Vega** (α Lyrae) and **Altair** (α Aquilae) form the striking Summer Triangle. The constellations of **Cygnus** (the Swan) and **Aquila** (the Eagle) represent birds 'flying' down the length of the Milky Way. This part of the Milky Way contains the Great Dark Rift, an elongated dark region where the light from distant stars is obscured by intervening dust. The dark Rift is clearly visible even to the naked eye in a dark sky region.

The most prominent stars of Cygnus are sometimes known as 'the Northern Cross' (as a counterpart to 'the Southern Cross' – the constellation of Crux – in the southern hemisphere). The central line of Cygnus through **Albireo** (β Cygni), extended well to the southwest, points to **Sabik** (η Ophiuchi) in the large, sprawling constellation of **Ophiuchus** (the Serpent Bearer) and beyond to **Antares** (α Scorpii) in the constellation of **Scorpius**. Like Cepheus, the shape of Ophiuchus somewhat resembles the gable end of a house, and the brightest star **Rasalhague** (α Ophiuchi) is at the 'apex' of the 'gable'.

A line from **Sadr** (γ Cygni) to Vega indicates the central portion, 'the Keystone', of the constellation of **Hercules**. **Draco** is adjacent on the eastern side of Hercules. An arc through the same stars, in the opposite direction, points towards the constellation of **Pegasus**, and more specifically to 'the Great Square of Pegasus'.

Aquila is less conspicuous than Cygnus and consists of a diamond shape of stars, representing the body and wings of the eagle, together with a rather faint star, λ Aquilae, marking the 'head'. **Lyra** (the Lyre) mainly consists of Vega (α Lyrae) and a small quadrilateral of stars to its southeast. Continuation of a line from Vega through Altair lands on the zodiacal constellation of **Capricornus**.

The Autumn Constellations

During the autumn season, the most striking feature is the Great Square of Pegasus, an almost perfect rectangle on the sky, forming the main body of the constellation of **Pegasus**. However, the star at the northeastern corner, **Alpheratz**, is actually α Andromedae, and part of the adjacent constellation of **Andromeda**. A line from **Scheat** (β Pegasi) at the northeastern corner of the Square, through **Matar** (η Pegasi), points in the general direction of Cygnus. A line from Markab through the last star in the Square, **Algenib** (γ Pegasi) points in the general direction of the five stars, including **Menkar** (α Ceti), that form the 'tail' of the constellation of **Cetus** (the Whale). A ring of seven stars lying below the southern side of the Great Square is known as 'the Circlet', part of the constellation of **Pisces** (the Fishes).

Extending the line of the western side of the Great Square towards the south leads to the isolated bright star, **Fomalhaut** (α Piscis Austrini), in the Southern Fish. Following the line of the eastern side of the Great Square towards the north leads to Cassiopeia while, in the other direction, it points towards **Diphda** (β Ceti), which is actually the brightest star in Cetus.

Three bright stars leading northeast from Alpheratz form the main body of the constellation of **Andromeda**. Continuation of that line leads towards the constellation of **Perseus** and **Mirfak** (α Persei). Running southwards from Mirfak is a chain of stars, one of which is the famous variable star **Algol** (β Persei), a star whose brightness changes periodically every 2.9 days. Farther east, an arc of stars leads to the prominent cluster of the **Pleiades**, in the constellation of **Taurus**.

Between Andromeda and Cetus lie the two small constellations of **Triangulum** (the Triangle) and **Aries** (the Ram).

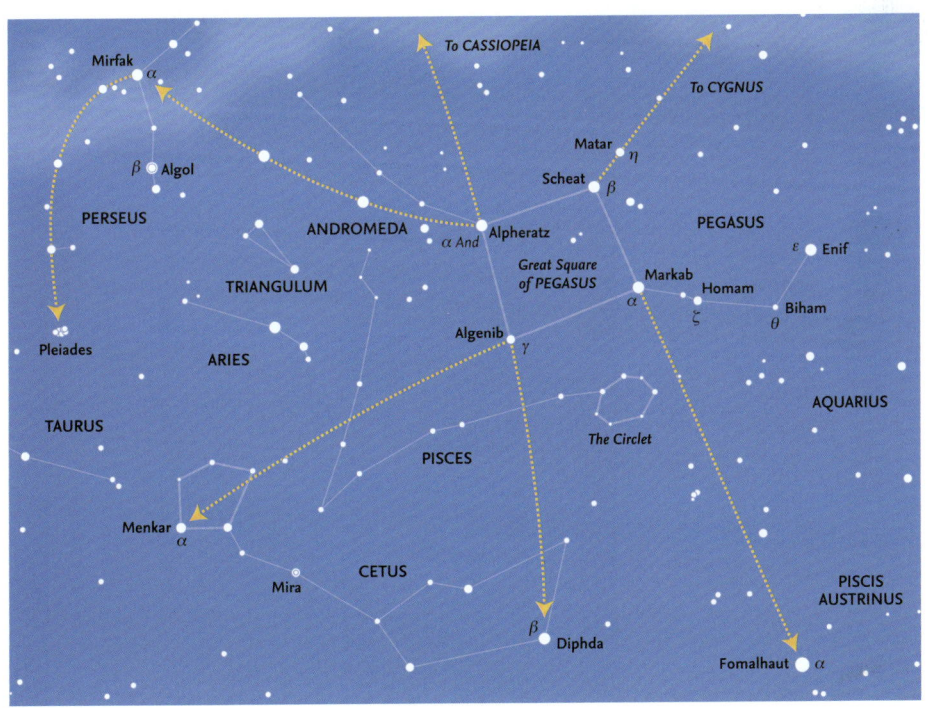

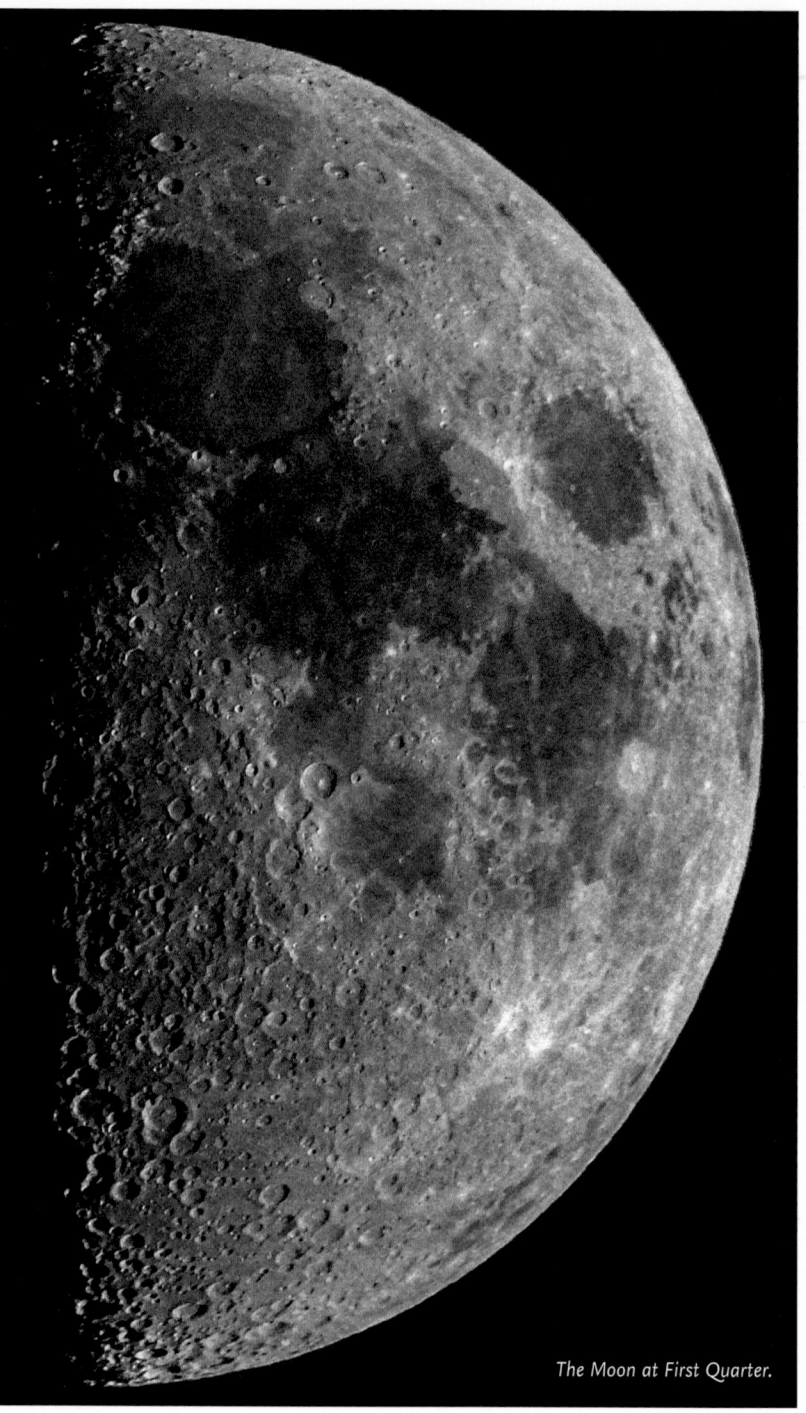

The Moon at First Quarter.

The Moon

The lunar phase cycle has a duration of 29.5 days and starts with the New Moon, when the near side of the Moon facing us is not illuminated by the Sun. The New Moon rises several hours after midnight and sets in the late afternoon or night; the First Quarter Moon rises in the late morning and sets around midnight; the Full Moon is above the horizon most of the night and the Last Quarter Moon rises around midnight and sets in the late morning.

Although the main features of the surface – the light highlands and the dark maria (seas) – may be seen with the naked eye, far more features may be detected with the use of binoculars or any telescope. The many craters are best seen when they are close to the **terminator** (the boundary between the illuminated and the non-illuminated areas of the surface), when the Sun rises or sets over any particular region of the Moon and the crater walls or central peaks cast strong shadows. These are best seen at crescent, quarter and gibbous phases. Most features become difficult to see at Full Moon, although this is the best time to see the bright ray systems surrounding certain craters. Accompanying the Moon map on the following pages is a list of prominent features, including the days in the lunation when they are normally close to the terminator and thus easiest to see. The dates of visibility vary slightly through the effects of **libration**. Because the Moon's orbit is inclined to the Earth's equator and also because it moves in an ellipse, the Moon appears to rock slightly from side to side (and nod up and down). Over time a total of 59 per cent of the lunar terrain can be observed. Areas near the **limb** (the edge of the Moon) may vary considerably in their location and visibility. This is easily noticeable with Mare Crisium (Sea of Crises, eastern limb of the Moon, in the western sky) and the craters Tycho (southern part of the Moon) and Plato (northern part). Another effect is that at crescent phases before and after New Moon, the normally non-illuminated portion of the Moon receives a certain amount of light, reflected from the Earth. This **Earthshine** may enable certain bright features (such as the craters Aristarchus, Kepler and Copernicus on the western limb of the Moon, in the eastern sky) to be detected.

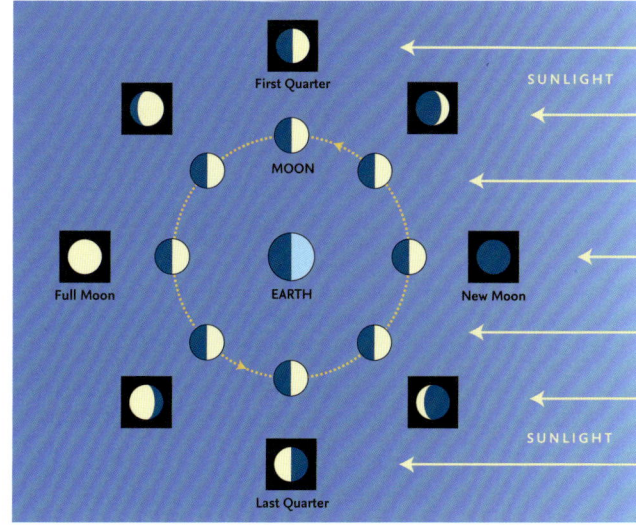

The Moon phases. *During its orbit around the Earth we see different portions of the illuminated side of the Moon's surface.*

Map of the Moon

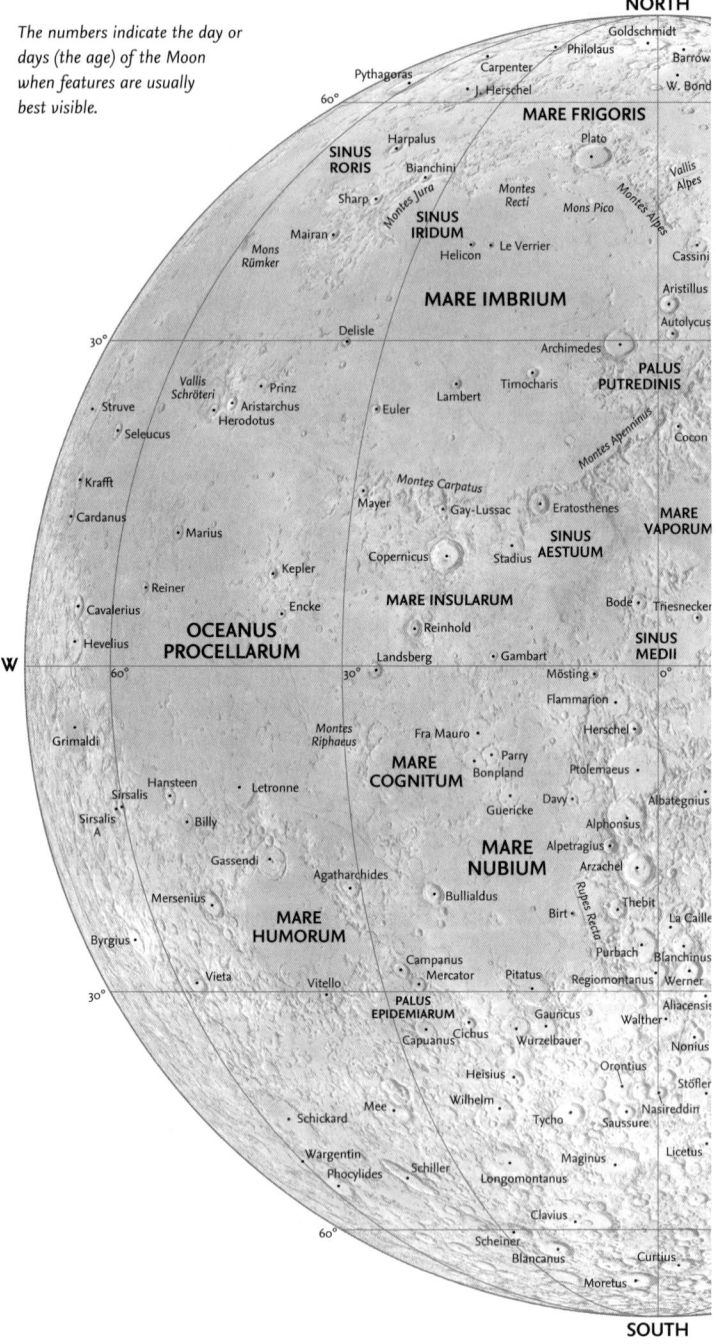

Abulfeda	6:20	*The numbers indicate the day or days (the age) of the Moon when features are usually best visible.*
Agrippa	7:21	
Albategnius	7:21	
Aliacensis	7:21	
Alphonsus	8:22	
Anaxagoras	9:23	
Anaximenes	11:25	
Archimedes	8:22	
Aristarchus	11:25	
Aristillus	7:21	
Aristoteles	6:20	
Arzachel	8:22	
Atlas	4:18	
Autolycus	7:21	
Barrow	7:21	
Billy	12:26	
Birt	8:22	
Blancanus	9:23	
Bullialdus	9:23	
Bürg	5:19	
Campanus	10:24	
Cassini	7:21	
Catharina	6:20	
Clavius	9:23	
Cleomedes	3:17	
Copernicus	9:23	
Cyrillus	6:20	
Delambre	6:20	
Deslandres	8:22	
Endymion	3:17	
Eratosthenes	8:22	
Eudoxus	6:20	
Fra Mauro	9:23	
Fracastorius	5:19	
Franklin	4:18	
Gassendi	11:25	
Geminus	3:17	
Goclenius	4:18	
Grimaldi	13-14:27-28	
Gutenberg	5:19	
Hercules	5:19	
Herodotus	11:25	
Hipparchus	7:21	
Hommel	5:19	
Humboldt	3:15	
Janssen	4:18	
Julius Caesar	6:20	
Kepler	10:24	
Landsberg	10:24	
Langrenus	3:17	
Letronne	11:25	
Linné	6	
Longomontanus	9:23	

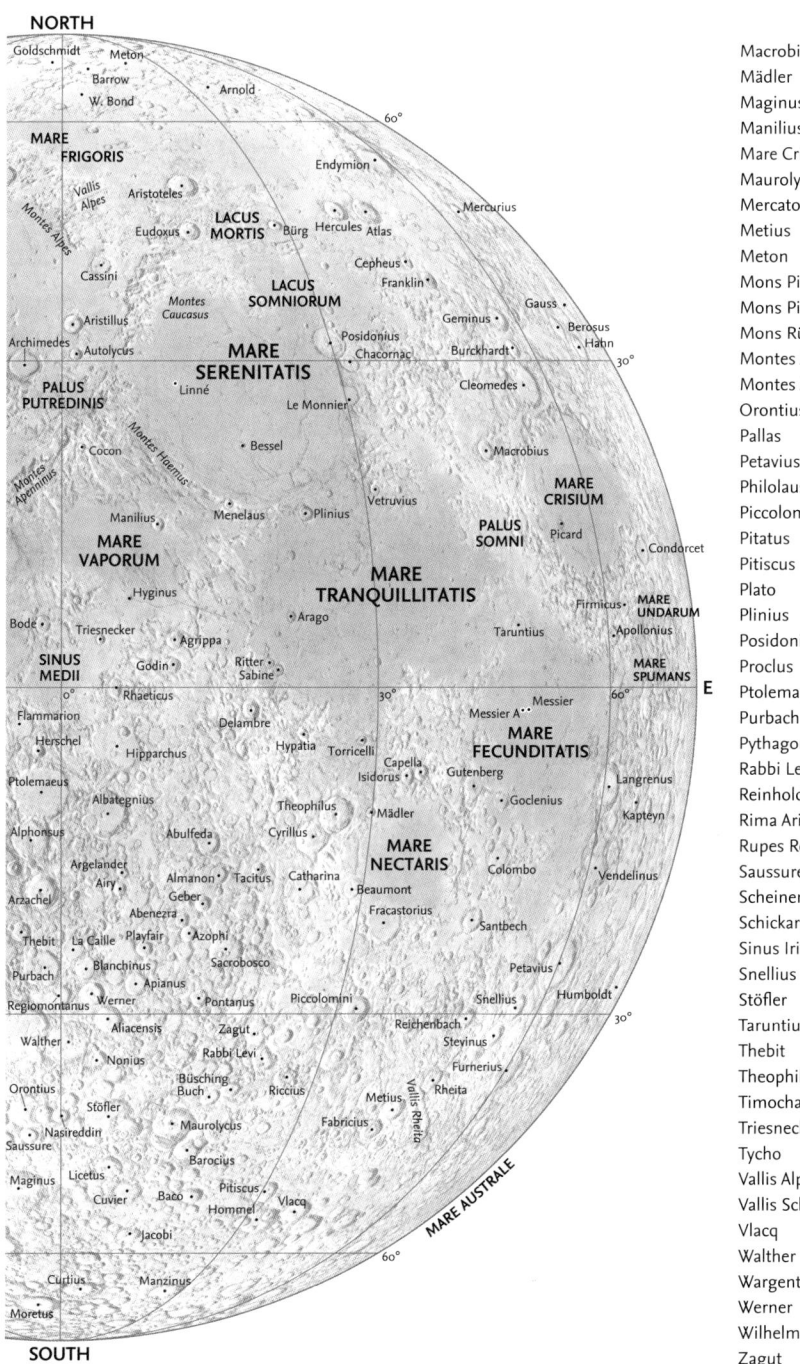

Macrobius	4:18
Mädler	5:19
Maginus	8:22
Manilius	7:21
Mare Crisium	2-3:16-17
Maurolycus	6:20
Mercator	10:24
Metius	4:18
Meton	6:20
Mons Pico	8:22
Mons Piton	8:22
Mons Rümker	12:26
Montes Alpes	6-8:21
Montes Apenninus	8
Orontius	8:22
Pallas	8:22
Petavius	3:17
Philolaus	9:23
Piccolomini	5:19
Pitatus	8:22
Pitiscus	5:19
Plato	8:22
Plinius	6:20
Posidonius	5:19
Proclus	14:18
Ptolemaeus	8:22
Purbach	8:22
Pythagoras	12:26
Rabbi Levi	6:20
Reinhold	9:23
Rima Ariadaeus	6:20
Rupes Recta	8
Saussure	8:22
Scheiner	10:24
Schickard	12:26
Sinus Iridum	10:24
Snellius	3:17
Stöfler	7:21
Taruntius	4:18
Thebit	8:22
Theophilus	5:19
Timocharis	8:22
Triesnecker	6-7:21
Tycho	8:22
Vallis Alpes	7:21
Vallis Schröteri	11:25
Vlacq	5:19
Walther	7:21
Wargentin	12:27
Werner	7:21
Wilhelm	9:23
Zagut	6:20

MAP OF THE MOON

Eclipses in 2026

Lunar eclipses
There are two lunar eclipses in 2026. The first is a total eclipse, which takes place on 3 March, and will be visible from eastern Asia and Australia as well as parts of North and South America. The second is a partial eclipse occurring on 28 August; this will be visible from Europe, Africa and the eastern Pacific. During a total lunar eclipse, the Moon moves eastwards into the Earth's penumbra, then moves behind the Earth into the darkest region of the Earth's shadow – the umbra – eventually creeping back out into the penumbra and into the sunlight. On 3 March maximum eclipse occurs at 11:34 and the Moon will spend a total of 58 minutes immersed in the darkest shadow of the Earth. During this time the Moon is still visible; however, it will appear a reddish-brown colour. This is due to refracted sunlight which passes through the Earth's atmosphere and bends towards the lunar surface. Bluer hues are scattered outwards by the Earth's atmosphere, leaving behind redder light that eventually reaches the Moon. On the early morning of 28 August, the time of maximum eclipse will be 04:13; the umbral (partial) eclipse will last 3 hours 18 minutes.

Solar eclipses
There are two solar eclipses in 2026. The first is an annular eclipse that takes place on 17 February with the maximum occurring at 12:12, visible from Antarctica; it will appear as a partial eclipse from south Argentina, Chile and South Africa. The duration of the annular eclipse is 2 minutes 20 seconds. There will be a total eclipse on 12 August visible from Iceland, Spain, Greenland and the Arctic. Maximum occurs at 17:46, the duration is 2 minutes 18 seconds. A partial solar eclipse will be visible from Europe, west Africa and North America. From London the duration of the eclipse is 1 hour 49 minutes.

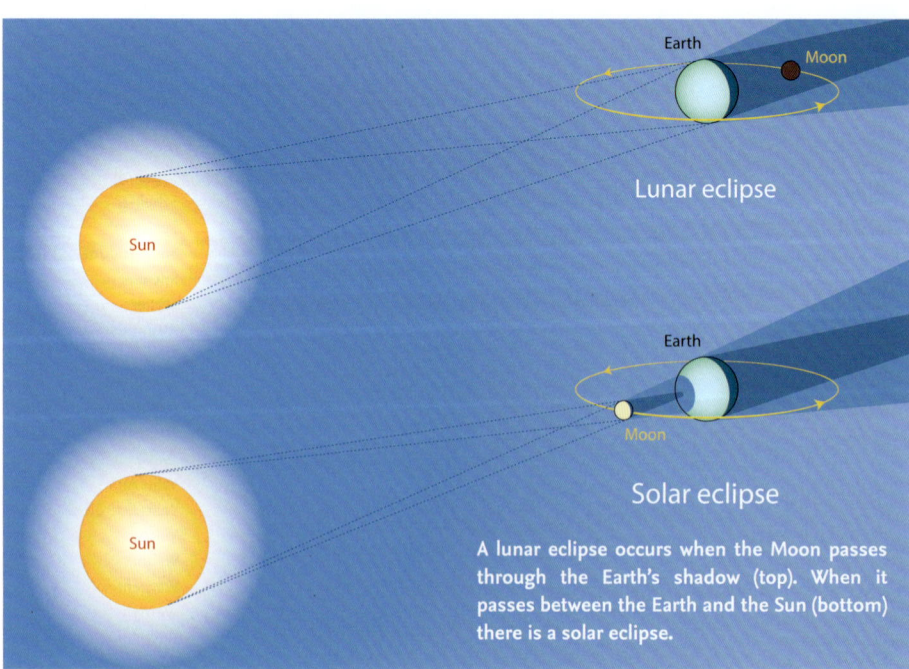

A lunar eclipse occurs when the Moon passes through the Earth's shadow (top). When it passes between the Earth and the Sun (bottom) there is a solar eclipse.

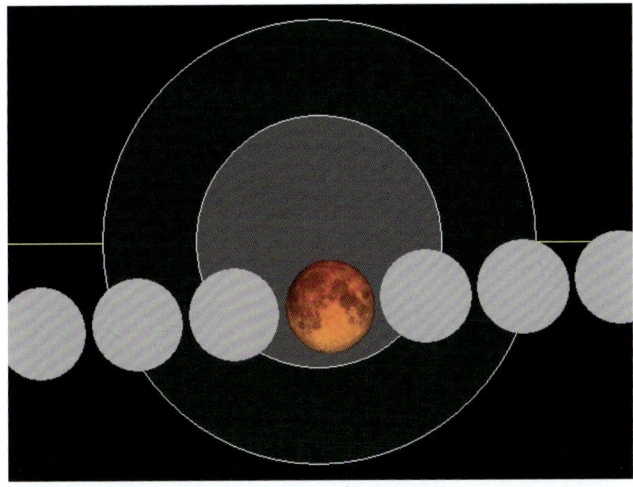

The path of the Moon as it pases through the shadow of the Earth on 3 March. Maximum eclipse occurs at 11:34 UT.

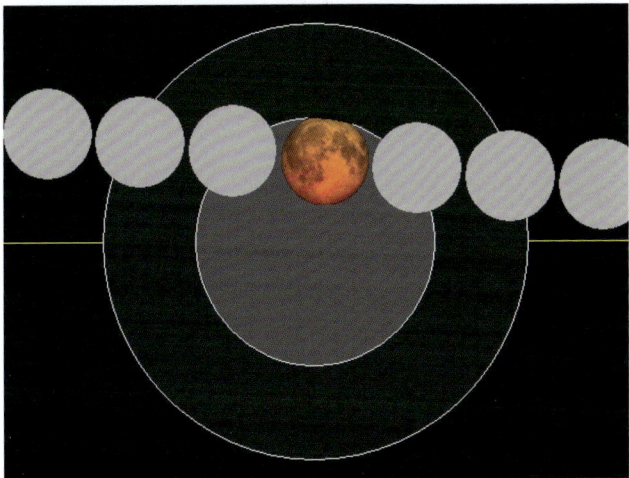

The path of the Moon as it passes through the shadow of the Earth on 28 August. Maximum eclipse occurs at 04:13 UT.

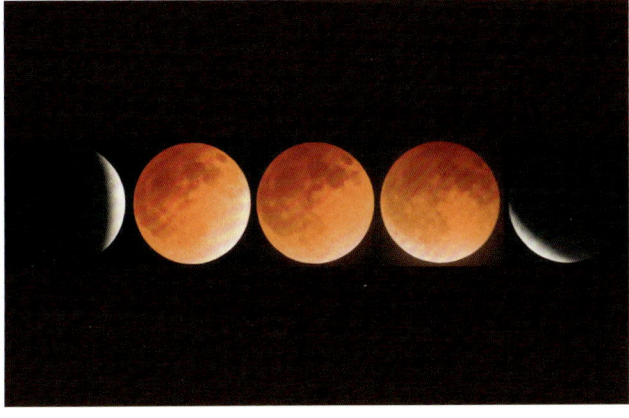

A sequence of five images showing the central stages of an eclipse, taken from Cascade-Siskiyou National Monument, Oregon in 2014.

ECLIPSES   21

The Planets in 2026

Mercury and Venus

Mercury reaches greatest eastern elongation three times during 2026 – it will be seen in the early evening of 19 February (Venus and Saturn are close by), 15 June (Venus and Jupiter are setting with Mercury) and 12 October, although it will be just above the horizon at sunset on this day. Mercury switches to an appearance at dawn when it is at greatest western elongation on 3 April, 2 August and 20 November (with Venus close by), although in April it will be positioned at a low altitude with Mars adjacent to the east. Its magnitude oscillates throughout the year – it will be brightest in May (-2.4) and faintest in November (6.5).

Venus moves from the dawn to the evening sky at the beginning of the year. It reaches greatest eastern elongation on 15 August appearing as the evening star, lying in Virgo. Venus leads the Sun in the dawn sky in late October, moving apart to reach greatest western elongation on 3 January 2027. Venus is brightest (mag. -4.9) in late November, it is dimmest (mag. -3.9) throughout most of March.

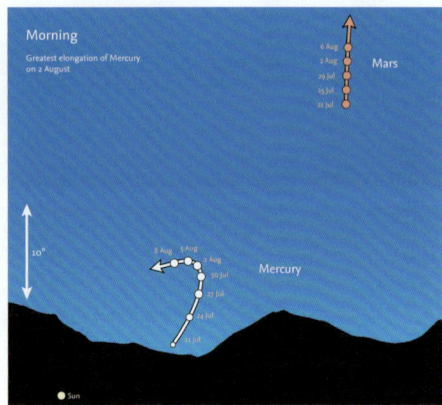

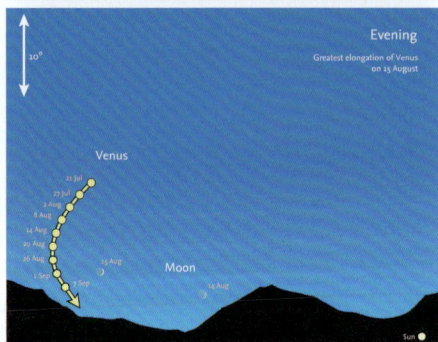

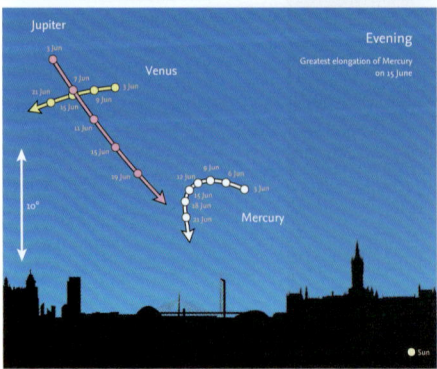

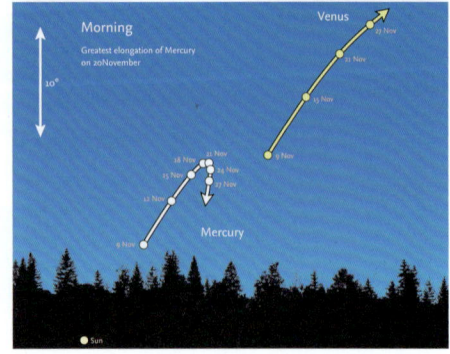

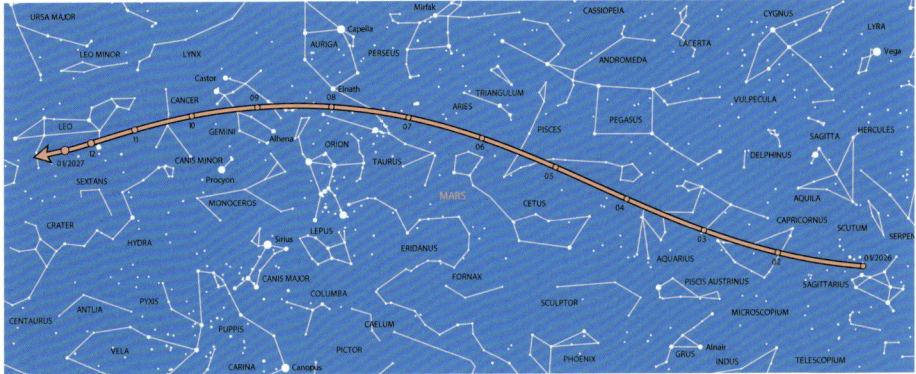

The path of Mars in 2026. Mars is visible at dawn for most of the year, eventually rising before midnight from November.

Mars

Mars is at its faintest in June (mag. 1.4) and it will brighten in the last half of the year, reaching a peak magnitude of -0.1 at the end of the year, when it will sit in **Leo**. Its apparent size in the sky will increase as it reaches opposition, which will next occur on 19 February 2027. Mars starts the year along the ecliptic in **Sagittarius**, moves into **Capricornus** in January and pulls away from the Sun in the sky, appearing at dawn by February. It progresses into **Aquarius**, **Pisces** in April and **Aries** in May. Mars will spend midsummer in **Taurus**, eventually moving into **Gemini** in August, **Cancer** in early Autumn and it will stay in **Leo** from Halloween onwards. Mars will rise earlier, creeping above the horizon a few hours before midnight by the end of the year.

As it takes longer to complete its orbit than the Earth, Mars does not come to opposition every year; the last opposition took place on 16 January 2025. Oppositions occur during a period of retrograde motion, when the planet appears to move westwards against the pattern of distant stars. Mars will continue to move eastwards until 10 January 2027 when it will appear to move backwards as the Earth overtakes it in its orbit around the Sun.

Because of its eccentric orbit, which carries it at differing distances from the Sun (and Earth), not all oppositions of Mars are equally favourable for observation. The relative positions of Mars and the Earth are shown below. It will be seen that the opposition of 2018 was very close and thus favourable for observation, and that of 2020 was also reasonably good. By comparison, opposition in 2027 will be at a far greater distance, so the planet will not appear as large as it did in 2018.

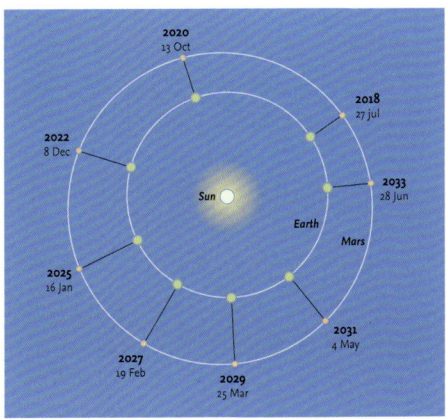

The oppositions of Mars between 2018 and 2033. The next opposition takes place 19 February 2027.

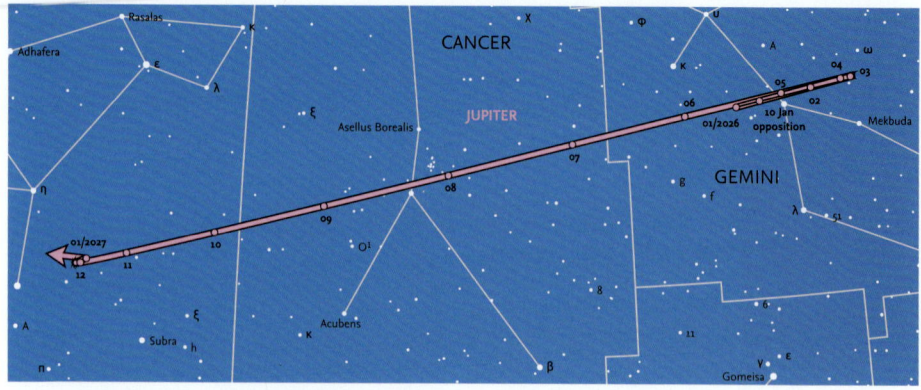

The path of Jupiter in 2026. Jupiter reaches opposition on 10 January. It ends retrograde motion on 11 March and re-enters westward motion on 13 December.

Jupiter and Saturn

Jupiter starts the year dazzlingly bright at magnitude -2.7. It reaches opposition on 10 January in Gemini, visible from sunset and moving westwards. Jupiter ends its retrograde motion on 11 March. It moves into Cancer in June and dims to mag. -1.8 in late July, lost in the glare of the Sun as it switches to the dawn sky. Jupiter crosses into Leo in September and it re-enters retrograde westward motion on 13 December at mag. -2.3.

Jupiter's four largest satellites are readily visible in binoculars. Not all four are visible all the time, they are sometimes hidden behind the planet or invisible in front of it. Io, the closest to Jupiter, orbits in just under 1.8 days, and Callisto, the farthest away, takes about 16.7 days. In between are Europa (c. 3.6 days) and the largest, Ganymede (c. 7.2 days).

Saturn is in *Aquarius* at the start of the year, it moves into *Pisces* by the end of January, appearing shortly after sunset at its faintest (mag. 1.0). The ringed planet swings into the dawn sky in late March and pulls away from the Sun, creeping above the horizon before midnight from August. On 26 July Saturn will enter retrograde motion. It will reach opposition on 4 October, and it will shine brightly with a magnitude of 0.3. This will be a good time to look at the planet although its rings will be close to edge-on, inclined by only 7° to our line of sight (allowing observers to see more of the southern side). Saturn will end its retrograde movement on 11 December.

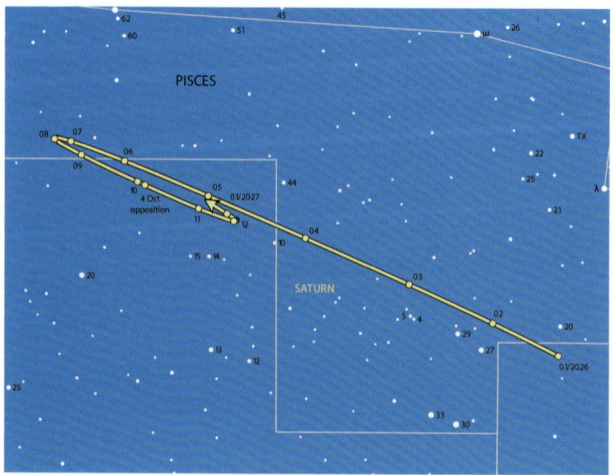

The path of Saturn in 2026. Saturn enters retrograde motion on 26 July, reverting back to direct motion on 11 December. It reaches opposition on 4 October.

Uranus and Neptune

Uranus sits in Taurus for the whole of 2026. It leads the Sun in the dawn sky in late May. It ends its retrograde motion on 4 February (it has been moving westwards since September 2025). It will re-enter retrograde motion on 10 September, moving westwards across the sky until 8 February 2027. It will reach opposition on 25 November (mag. 5.6).

Neptune moved into **Pisces** in May 2022; it will stay in Pisces until August 2039. In March it crosses from the evening to the dawn sky, sitting low in the east before sunrise for the rest of Spring. It enters retrograde motion on 7 July at mag. 7.9, visible after midnight. It will reach opposition on 26 September (mag. 7.7). It reverts to direct motion on 12 December and will continue moving eastwards until 10 July 2027.

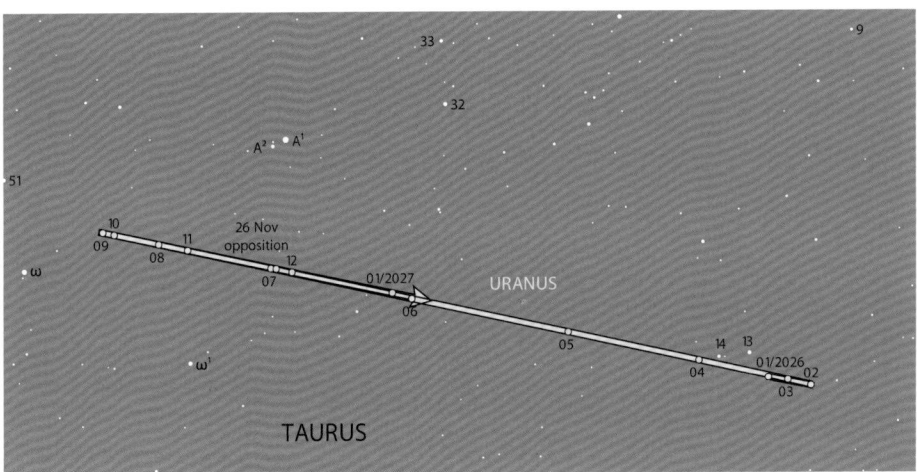

The path of Uranus in 2026 in Taurus. Uranus ends retrograde motion on 4 February, re-entering on 10 September. It reaches opposition on 25 November.

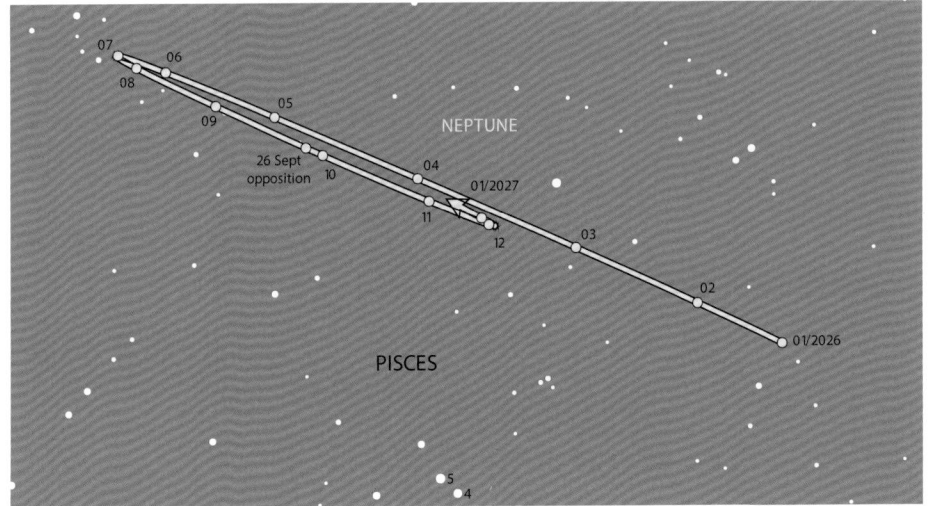

The path of Neptune in 2026 in Pisces. Pisces enters retrograde motion on 7 July, reaching opposition on 26 September. It reverts to direct eastward motion on 12 December.

THE PLANETS   **25**

Minor Planets in 2026

Three minor planets come to opposition and rise above magnitude 9.0 in 2026. They are closest to Earth at this point and reach their highest point in the sky around midnight – this is the best time to see them. On 30 September the asteroid *(192) Nausikaa* will lie in **Pisces**, reaching a peak magnitude of 8.4 at opposition. On 4 October *(2) Pallas* will reach magnitude 8.2 in **Cetus** and *(4) Vesta* will reach opposition later in the month on 13 October, visible at magnitude 6.5.

In December the dwarf planet *(1) Ceres* will increase in brightness from magnitude 7.6 to 6.9, reaching opposition on 7 January 2027 at a distance of 1.63 AU (Astronomical Units, equivalent to the average distance from the Earth to the Sun). Ceres is the only dwarf planet that lies within the orbit of Neptune (it is part of the asteroid belt between Mars and Jupiter) and it completes its orbit in 4.6 years. The Earth could accommodate 2,500 of Ceres.

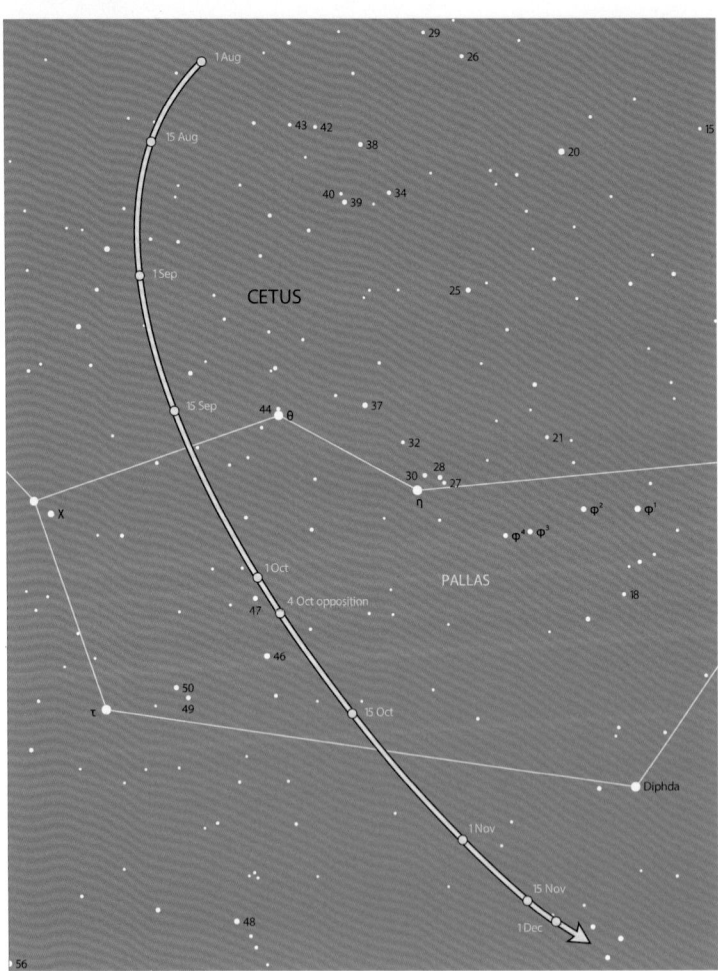

The path of (2) Pallas in 2026. (2) Pallas reaches opposition on 4 October 2026.

MINOR PLANETS 27

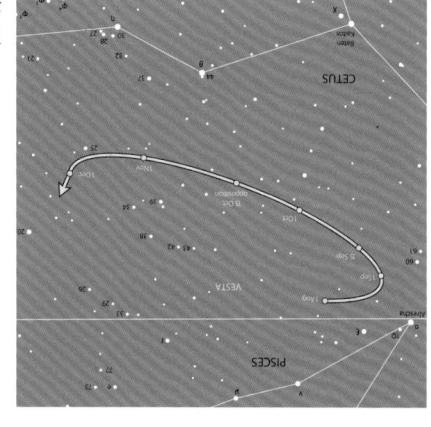

The path of (4) Vesta in 2026. (4) Vesta reaches opposition on 13 October.

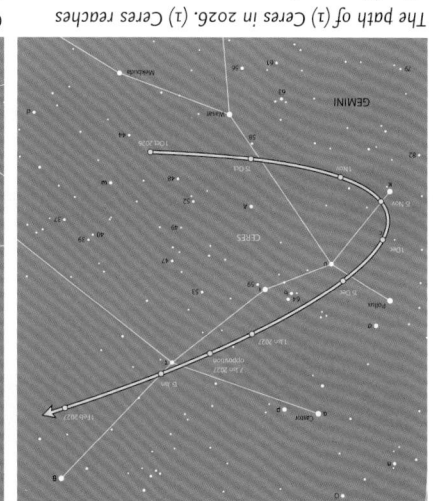

The path of (1) Ceres in 2026. (1) Ceres reaches opposition on 7 January 2027.

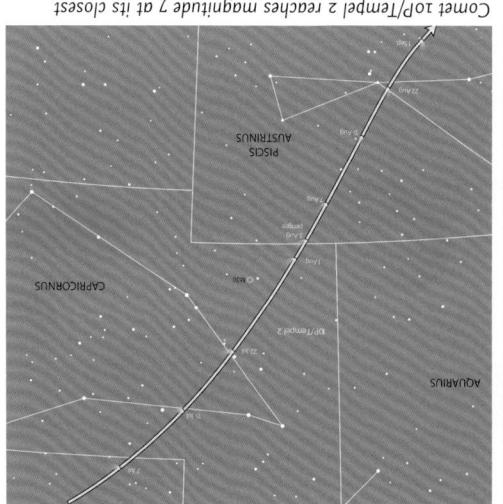

Comet 10P/Tempel 2 reaches magnitude 7 at its closest approach to Earth on 3 August.

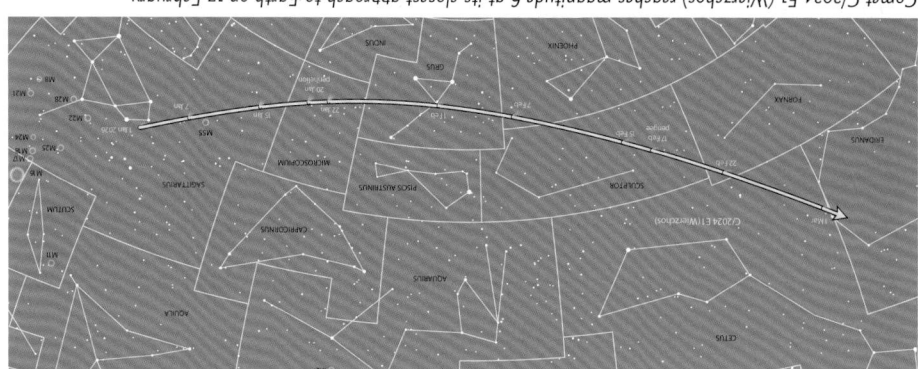

Comet C/2024 E1 (Wierzchoś) reaches magnitude 6 at its closest approach to Earth on 17 February.

Comets in 2026

Although comets may occasionally become very striking objects in the sky, their occurrence and particularly the existence or length of any tail and their overall magnitude are notoriously difficult to predict. Naturally, it is only possible to predict the return of periodic comets (whose names have the prefix 'P'). Many comets appear unexpectedly (these have names with the prefix 'C'). Bright, readily visible comets such as C/1995 Y1 Hyakutake, C/1995 O1 Hale-Bopp, C/2006 P1 McNaught or C/2020 F3 NEOWISE are rare. Most periodic comets are faint and only a very small number ever become bright enough to be easily visible with the naked eye or with binoculars.

Comet 24P/Schaumasse will be visible most of the night from January to March, brightening from magnitude 9 to 8, moving through the constellations of Virgo, Boötes and Serpens Caput. It will make its closest approach on 4 January 2026, passing within a distance equivalent to 60% of the Earth-Sun distance. Comet C/2024 E1 (Wierzchos) will pop into Cetus, Eridanus and Taurus in early Spring, brightening to magnitude 6 at its closest point to Earth on 17 February. Comet 10P/Tempel 2 appears in June in Aquila at mag. 9, visible after 23:00. It moves into Capricornus in July and August, brightening to mag. 7 at point of closest approach on 3 August. Comet 10P/Tempel 2 has a period of 5.4 years.

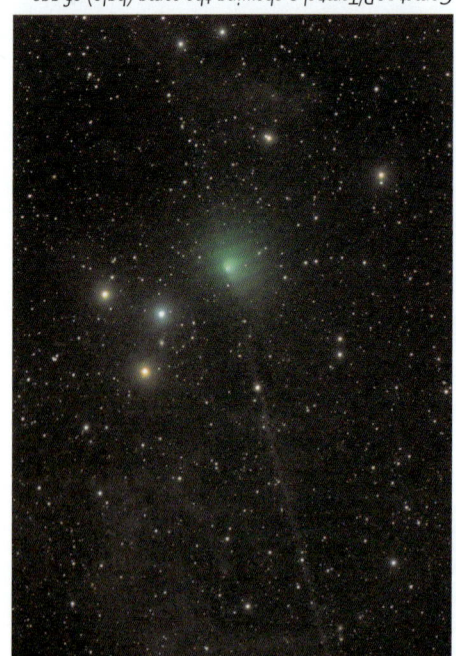

Comet 10P/Tempel 2 showing the coma (halo) of gas surrounding the nucleus.

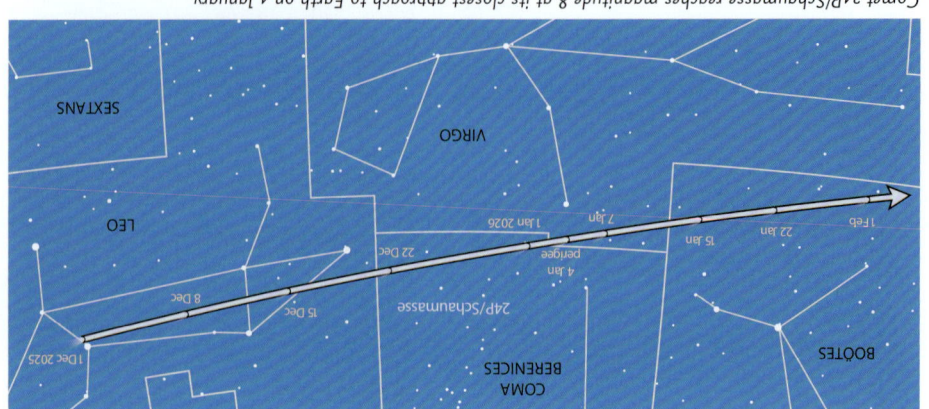

Comet 24P/Schaumasse reaches magnitude 8 at its closest approach to Earth on 4 January.

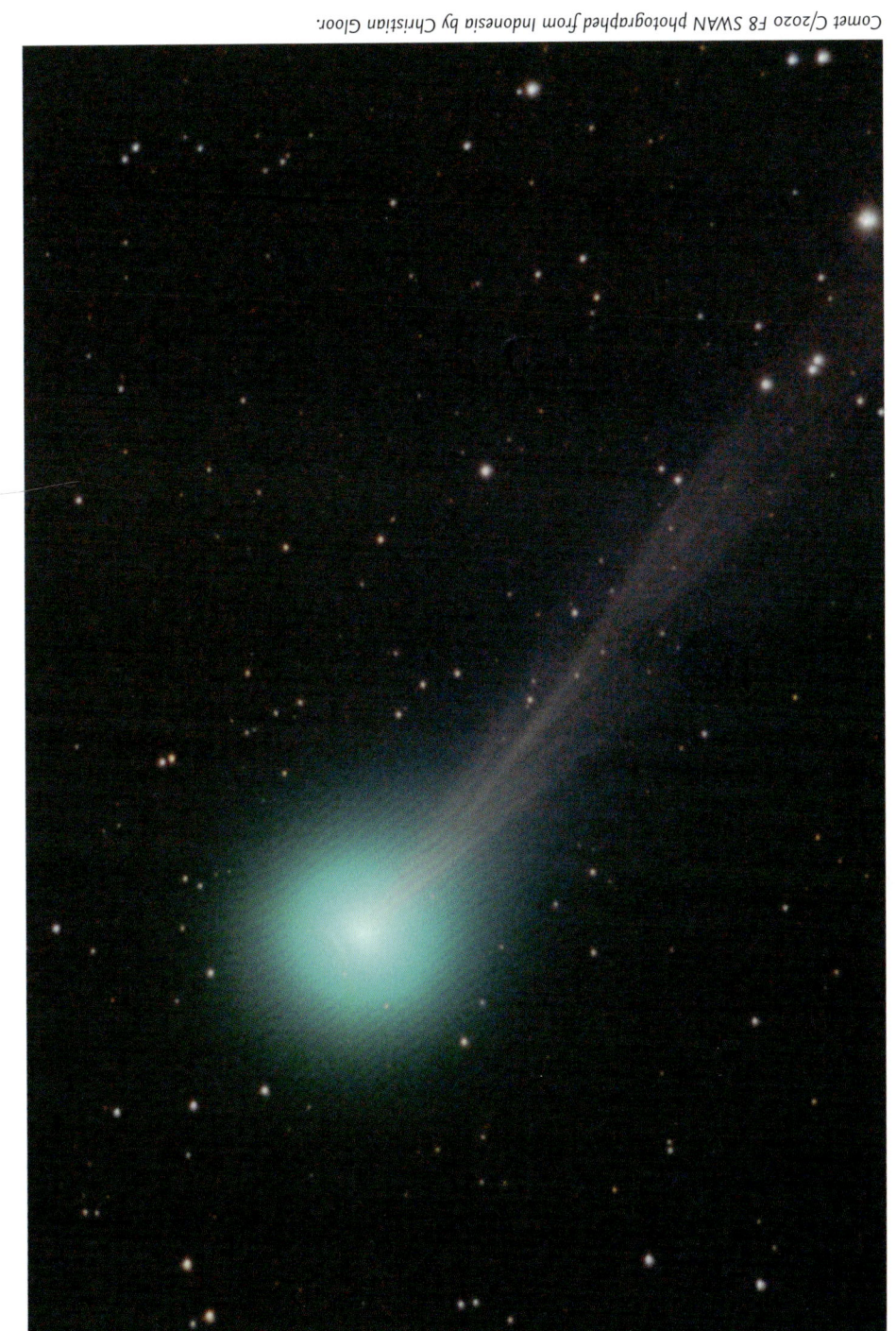

Comet C/2020 F8 SWAN photographed from Indonesia by Christian Gloor.

Introduction to the Month-by-Month Guide

The monthly charts

The pages devoted to each month contain a pair of charts showing the appearance of the night sky, looking north and looking south. The charts are drawn for the latitude of 50°N, so observers farther north will see slightly more of the sky on the northern horizon, and slightly less on the southern. Stars close to the horizon are always dimmed by atmospheric absorption (this is called extinction), so sometimes the faintest stars marked on the charts may not be visible.

The charts are drawn to show the appearance at 23:00 GMT for the 1st of each month. The same appearance will apply an hour earlier (22:00 GMT) on the 15th, and yet another hour earlier (21:00 GMT) at the end of the month (shown as the 1st of the following month). GMT is identical to the Universal Time (UT) used by astronomers around the world. In Europe, Summer Time is introduced in March, so the March charts apply to 23:00 GMT on 1 March, 22:00 GMT on 15 March, but 22:00 BST (British Summer Time) on 1 April. The change back from Summer Time (in Europe) occurs in October, so the charts for that month apply to 00:00 BST for 1 October, 23:00 BST for 15 October and 21:00 GMT for 1 November. The charts may be used for earlier or later times during the night. To observe two hours earlier, use the charts for the preceding month; for two hours later, the charts for the next month.

Tips for observing the night sky

A successful observing session requires planning, flexibility and patience. Check the weather forecast, sunset and moonrise and moonset times and choose a location where you have a clear view of the horizon wherever possible. Bring your star guide or app on a mobile device, a red torch to aid your night vision and a pair of binoculars and a tripod to keep them steady. You may be outside for a while – it's advised that you wear warm clothes, socks and shoes and bring hot drinks if it's a cold night.

If you are planning on looking for meteors or looking up at the night sky for a while you may wish to bring a deckchair or picnic blanket with you. When looking at faint objects (unlike the Moon) you will need to give your eyes 10–20 minutes to adapt to the dark and achieve night vision. If you are using your

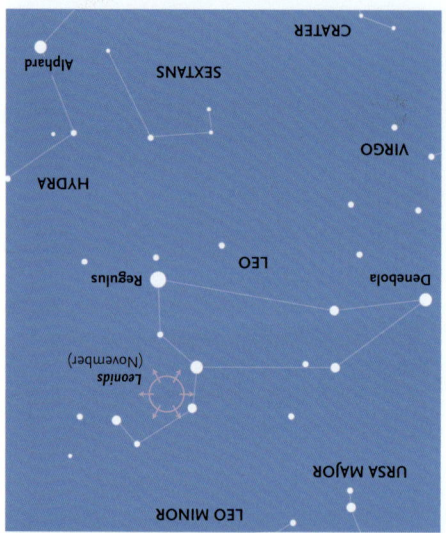

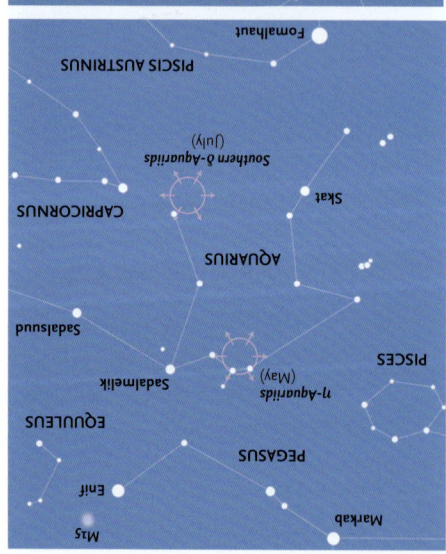

30

THE MONTH-BY-MONTH GUIDE

phone, switch it to night vision mode and use a red torch to find your way around so that you maintain your night vision.

Astrophotography

You can capture the Moon and five planets visible to the naked eye or the brightest stars using your smartphone. The challenge is keeping your phone steady and gaining control over the focus and exposure (if the camera allows it). It is advised to attach the phone to a tripod, or you could use an adaptor to attach your phone to the eyepiece of a telescope – this will allow you to align it correctly with minimal effort. You can use the zoom function on your camera to change the magnification of the image.

There are many apps available that will provide more control of your camera settings and allow you to take better quality images: NightCap Pro, ProCam and AstroShader for iOS and Open Camera, DeepSkyCamera or Camera FV-5 for Android. For the Moon and bright planets, your camera should auto lock and focus, otherwise manually set the focus to infinity and keep the exposure short (a fraction of a second if it is a gibbous or Full Moon, one or two seconds for a thin crescent Moon and planets). You could control the shutter remotely via the volume control on wireless headphones, or set a timer on your phone (3 seconds delay). For fainter objects, increase the exposure and the ISO (the sensitivity of the camera detector); however, a higher ISO setting will introduce more noise. Trial and error will help you choose the best settings. Some apps allow images to be stacked; this means you can overlay images to increase brightness further. This is useful for the constellations, clusters such as the Pleiades and nebulous objects such as the Orion Nebula and Andromeda.

Meteors

Details of specific meteor showers are given in the months when they come to maximum. Radiants that lie below the horizon or in constellations not readily visible in that month are not marked. For this reason, special charts for the Eta (η) and Delta (δ) Aquariids (May and July, respectively) and the Leonids (November) are given here. Meteors from such showers may still be seen, because the most effective region for seeing meteors is some 40–45° from the radiant, and that area of sky may well be above the horizon. A table of the best meteor showers visible during the year is also given here. The rates given are those that an experienced observer might see under ideal conditions. Generally, the observed rates will be far lower.

The photographs

One or more photographs of constellations visible in certain specific months are included.

Shower	Dates of activity 2026	Date of maximum 2026	Possible hourly rate
Quadrantids	28 December to 12 January	4 January	120
April Lyrids	16 April to 25 April	22 April	18
η-Aquariids	19 April to 28 May	6 May	40
α-Capricornids	3 July to 15 August	30 July	5
Southern δ-Aquariids	12 July to 23 August	30 July	25
Perseids	17 July to 24 August	13 August	150
α-Aurigids	28 August to 5 September	1 September	6
Southern Taurids	10 September to 20 November	10 October	5
Orionids	2 October to 7 November	21 October	15
Draconids	6 October to 10 October	9 October	var.
Northern Taurids	20 October to 10 December	12 November	5
Leonids	6 November to 30 November	18 November	15
Geminids	4 December to 20 December	14 December	120
Ursids	17 December to 26 December	22 December	10

The Moon

The section on the Moon includes details of any lunar or solar eclipses that may occur during the month. Similar information is given about any important occultations. Mainly, however, this section summarizes when the Moon passes close to planets or the five prominent stars close to the ecliptic. The dates when the Moon is closest to the Earth (at *perigee*) and farthest from it (at *apogee*) are shown in the monthly calendars, and only the nearest and farthest points during the year are mentioned.

The planets and minor planets

Brief details are given of the location, movement and brightness of the planets from month to month. Mercury and Venus are normally easiest to see around either eastern or western elongation, in the evening or morning sky, respectively. The dates at which the superior planets reverse their motion (from direct motion to *retrograde*, and retrograde to direct) and of opposition (when a planet generally reaches its maximum brightness) are given. Jupiter and Saturn are normally easiest to see around opposition, which occurs every year. Mars, by contrast, moves relatively rapidly against the background stars and in some years never comes to opposition. The distant outer planets, especially Saturn, Uranus and Neptune, may spend most or all of the year in a single constellation.

Uranus is normally magnitude 5.7 to 5.9, and thus at the limit of naked-eye visibility under exceptionally dark skies, but bright enough to be readily visible in binoculars. Similar considerations apply to Neptune; it is always fainter (generally magnitude 7.8 to 8.0); however, it is still visible in most binoculars. The charts on page 25 show their positions during 2026.

In any year, few minor planets ever become bright enough to be detectable in binoculars. Just one, (4) Vesta, on rare occasions brightens sufficiently for it to be visible to the naked eye. Our limit for visibility is magnitude 9.0 and details and charts are given for those objects that exceed that magnitude during the year.

The Moon calendar

The Moon calendar shows the phase of the Moon for every day of the month. Also shown is the *age* of the Moon (the day in the *lunation*), beginning at New Moon, which may be used to determine the best time for observation of specific lunar features.

It should be noted, however, that several factors affect observation: differences between the sensitivity of different individuals to faint starlight including colour, the degree to which they have become adapted to the dark, the local altitude of the star and the atmospheric conditions. Also, the exposure time can be altered on a digital camera, thus enhancing the intensity and colour of the object.

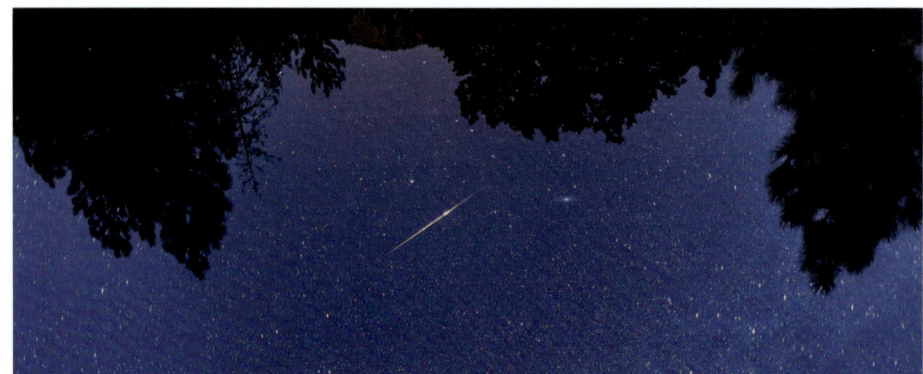

A Perseid fireball over Oregon.

THE MONTH-BY-MONTH GUIDE

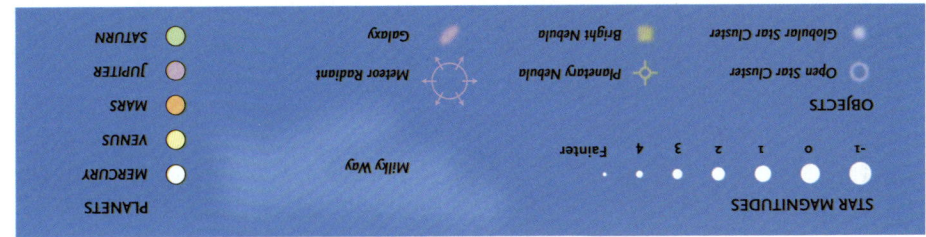

Key to the symbols used on the monthly star maps.

The ecliptic charts

Although the ecliptic charts are primarily designed to show the positions and motions of the major planets, they also show the motion of the Sun during the month. The light-tinted area shows the area of the sky that is invisible during daylight, but the darker area gives an indication of which constellations are likely to be visible at some time of the night.

The monthly calendar

For each month, a calendar shows details of significant events, including when planets are close to one another in the sky, close to the Moon, or close to any one of five bright stars that are spaced along the ecliptic. The times shown are given in Universal Time (UT), which is identical to Greenwich Mean Time (GMT).

The diagrams of interesting events

Each month, a number of diagrams show the appearance of the sky when certain events take place. However, the exact positions of celestial objects and their separations greatly depend on the observer's position on Earth. When the Moon is one of the objects involved, because it is relatively close to Earth, there may be very significant changes from one location to another. Close approaches between planets or between a planet and a star are less affected by changes of location, which may thus be ignored. The diagrams showing the appearance of the sky are drawn for the latitude of London, so

normally around opposition. The background stars are shown to a fainter magnitude for better contrast. Minor planet charts for 2026 are on pages 26 and 27.

will be approximately correct for most of Britain and Europe. However, for an observer farther north (say, in Edinburgh), a planet or star listed as being north of the Moon will appear even farther north, whereas one south of the Moon will appear closer to it – or may even be hidden (occulted) by it. For an observer farther south than London, there will be corresponding changes in the opposite direction: for a star or planet south of the Moon, the separation will increase, and for one north of the Moon, the separation will decrease. This is particularly important when occultations occur, which may be visible from one location, but not another.

The details given regarding the positions of the various bodies should be used as a guide to their location. A similar situation arises with the times that are shown. Similarly, dates and times are given, even if they fall in daylight, when the objects are likely to be completely invisible. However, such times do give an indication that the objects concerned will be in the same general area of the sky during both the preceding and the following nights.

Data used in this Guide

The data given in this Guide, such as timings and distances between objects, have been determined by the following sources: the Astronomical Applications Department of the US Naval Observatory; the Sky Events Calendar by Fred Espenak and Sumit Dutta (NASA's Goddard Space Flight Center); the NASA Jet Propulsion Laboratory (JPL) Horizons System; the International Astronomical Union's (IAU) Minor Planet Center, In-The-Sky.org.

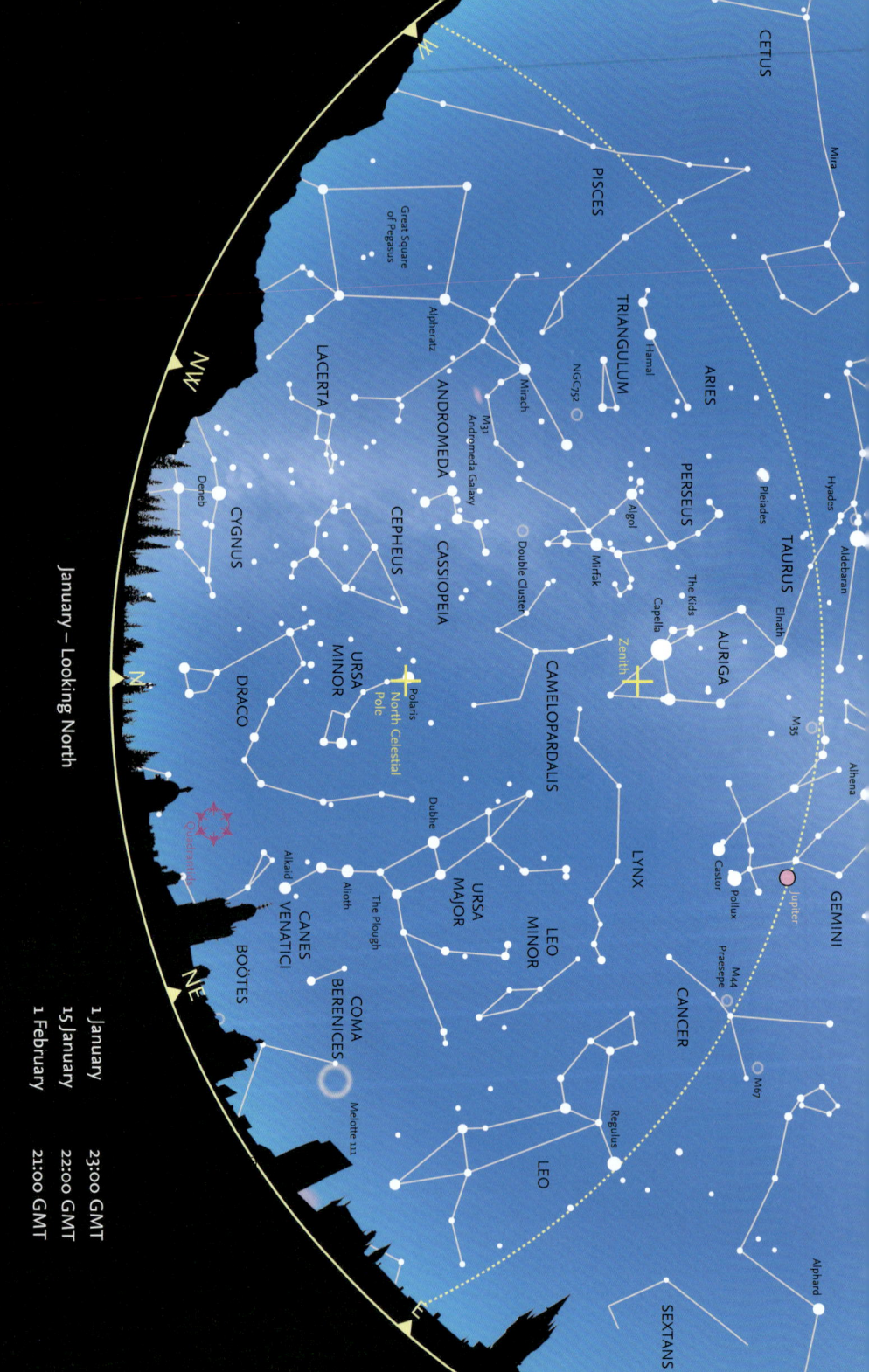

January – Looking North

Most of the important circumpolar constellations are easy to see in the northern sky at this time of year. **Ursa Major** stands more or less vertically above the horizon in the northeast, with the zodiacal constellation of **Leo** rising in the east. To the north, the stars of **Ursa Minor** lie below **Polaris** (the Pole Star). The head of **Draco** is low on the northern horizon, the stars may be difficult to see unless observing conditions are good. Both **Cepheus** and **Cassiopeia** are readily visible in the northwest, and even the faint constellation of **Camelopardalis** is high enough in the sky for it to be easily visible.

Near the zenith is the constellation of **Auriga** (the Charioteer), with brilliant **Capella** (α Aurigae), directly overhead. Slightly to the west of Capella lies a small triangle of fainter stars, known as the Kids. (Ancient mythological representations of Auriga show him carrying two young goats.) Together with the northernmost bright star in **Taurus**, **Elnath** (β Tauri), the body of Auriga forms a large pentagon on the sky, with the Kids lying on the western side. Farther down towards the west are the constellations of **Perseus** and **Andromeda**, and the Great Square of **Pegasus** is approaching the horizon.

Meteors

The **Quadrantid** meteor shower begins on 28 December 2025 and continues until 12 January 2026. It is one of the strongest and most consistent meteor showers. It comes to maximum on the night of 3 and 4 January, during a Full Moon, so moonlight will reduce visibility. The meteors are bright, bluish- or yellowish-white, and may reach a maximum rate of 110 per hour. The parent object is minor planet 2003 EH$_1$.

The shower is named after the former constellation **Quadrans Muralis** (the Mural Quadrant), an early form of astronomical instrument. The Quadrantid radiant, marked on the chart, is now within the northernmost part of **Boötes**, roughly halfway between θ Boötis and τ Herculis.

Ursa Major and the asterism, the Plough.

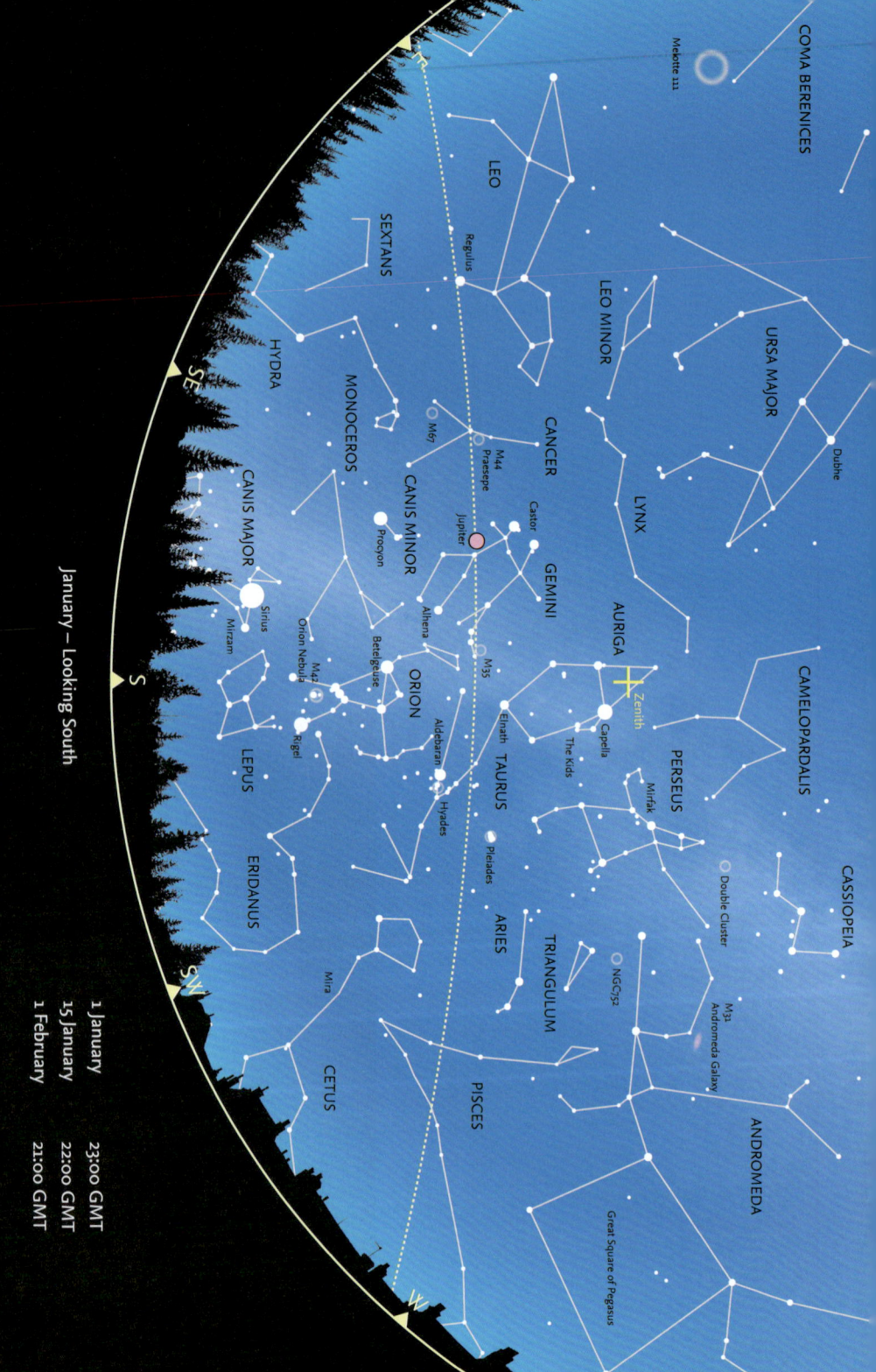

January – Looking South

The southern sky is dominated by **Orion**, prominent during the winter months, visible at some time during the night. It is highly distinctive, with a line of three stars that form Orion's Belt. To most observers, the bright star **Betelgeuse** (α Orionis), shows a reddish tinge; it is known as a red supergiant star, in contrast to the brilliant bluish-white **Rigel** (β Orionis). The three stars of the belt lie directly south of the celestial equator. A vertical line of three 'stars' forms the 'sword' that hangs south of the 'belt'. With good viewing, the central 'star' appears as a hazy spot, even to the naked eye, and is actually the **Orion Nebula**, a star formation region. Binoculars reveal the four stars of the **Trapezium** cluster, which illuminate the nebula.

The Belt points up to the northwest towards **Taurus** (the Bull) and orange-tinted **Aldebaran** (α Tauri). Close to Aldebaran, there is a conspicuous 'V' of stars, called the **Hyades** cluster. Aldebaran lies much closer to the Earth than the Hyades, it just so happens to share the same line of sight. Farther along, the same line from Orion passes below a bright cluster of stars, the **Pleiades**, or Seven Sisters. Even the smallest pair of binoculars reveals this as a beautiful group of bluish-white stars.

Another conspicuous star in Taurus, **Elnath** (β Tauri), lies directly above Orion and forms a big 'V' with Aldebaran.

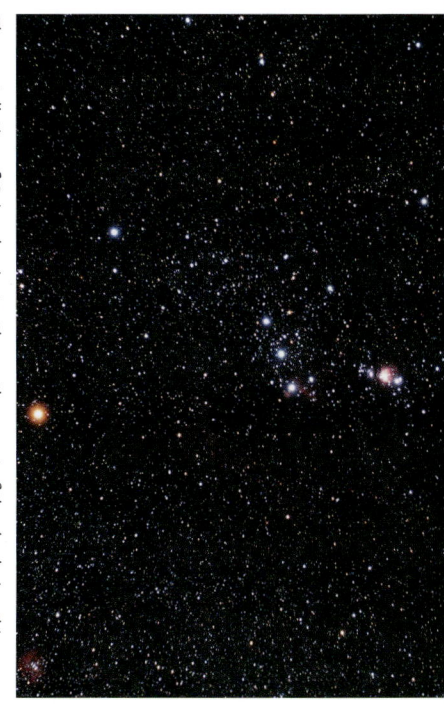

The constellation of Orion dominates the southern part of the sky during this period of the year, the three stars marking the belt are easy to find. Here, orange Betelgeuse, blue-white Rigel and the pink Orion Nebula are prominent.

The Moon's phases for January 2026

| 01 Day 13 | 02 Day 14 New Moon | 03 Day 15 | 04 Day 16 | 05 Day 17 | 06 Day 18 | 07 Day 19 | 08 Day 20 | 09 Day 21 | 10 Day 22 First Quarter | 11 Day 23 | 12 Day 24 | 13 Day 25 | 14 Day 26 | 15 Day 27 | 16 Day 28 |

| 17 Day 29 | 18 Day 30 | 19 Day 1 | 20 Day 2 Full Moon | 21 Day 3 | 22 Day 4 | 23 Day 5 | 24 Day 6 | 25 Day 7 | 26 Day 8 | 27 Day 9 Last Quarter | 28 Day 10 | 29 Day 11 | 30 Day 12 | 31 Day 13 |

January – Moon and Planets

The Moon

The Moon is Full on 3 January, **Jupiter** will appear 3.7°S of the Moon at mag. -2.7, both are visible from the early evening. A day later **Pollux** will be 3.0°N of the waning gibbous Moon. On 6 January **Regulus** lies only 0.5°S of the Moon. On 10 January **Spica** lies 1.6°N of the Last Quarter Moon; four days later the red-orange star **Antares** will be placed 0.6°N of the waning crescent Moon, appearing low in the east a few hours before sunrise. The lunar cycle resets on 18 January with a New Moon. On 23 January **Saturn** (mag. 1.0) will appear 4.3°S of the waxing crescent Moon in the early evening, both setting shortly after 20:30. Three days later the Moon will be First Quarter, a day later the **Pleiades** cluster lies 1.1°S of the Moon. The month ends with **Jupiter** (mag. -2.6) 3.8°S of the waxing gibbous Moon and **Pollux** lies 3.0°N.

The Planets

Mercury starts the month visible just before sunrise and trails the Sun in the evening after 21 January, brightening to mag. -1.3. **Venus** leads the Sun at dawn at the start of the year and transitions to the early evening sky after 6 January, visible at the end of the month at mag. -3.9. **Mars** approaches the Sun in Sagittarius and swings into the dawn sky; however, it is too close to the Sun to be seen (mag. 1.1 to 1.0, then dimming to 1.2). **Jupiter** lies in **Gemini**, it is best seen at opposition at midnight on 10 January, mag. -2.7. **Saturn** moves from **Aquarius** to **Pisces** during the month (mag. 1.0), **Neptune** is nearby. **Uranus** sits in **Taurus** until a few hours after midnight, at mag. 5.6. **Neptune** is in **Pisces** at mag. 7.8.

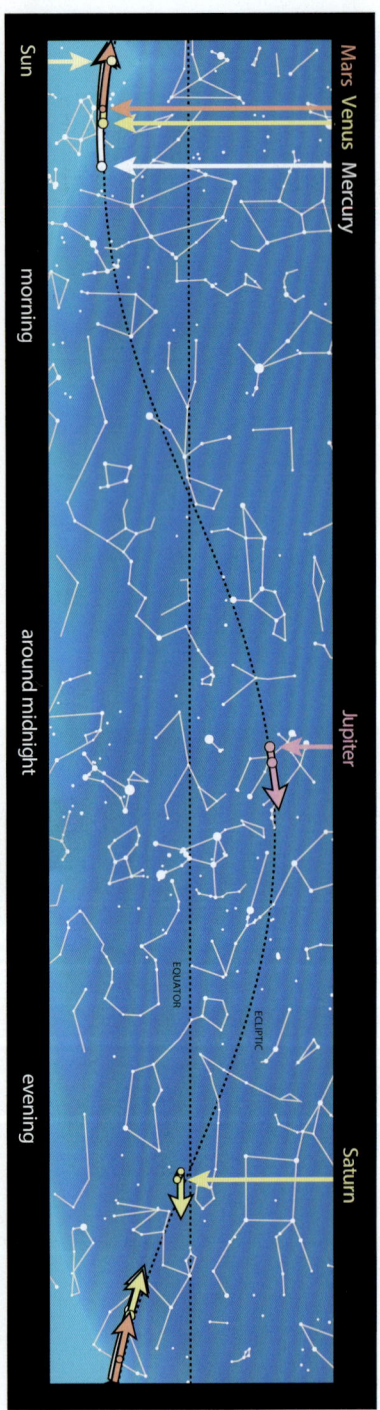

The path of the Sun and the planets along the ecliptic in January.

Calendar for January

01	21:43	Moon at perigee = 360,348 km
03	10:03	Full Moon
03	17:35	Earth at perihelion (147,099,900 km = 0.9833 AU)
03–04		Quadrantid meteor shower maximum
03	22:01	Jupiter (mag. -2.7) 3.7°S of the Moon
04	03:28	Pollux 3.0°N of the Moon
06	16:20	Regulus 0.5°S of the Moon
10	08:34	Jupiter (mag. -2.7) at opposition
10	15:48	Last Quarter Moon
10	23:50	Spica 1.6°N of the Moon
13	20:48	Moon at apogee = 405,437 km
14	19:28	Antares 0.6°N of the Moon
18	19:52	New Moon
23	12:31	Saturn (mag. 1.0) 4.3°S of the Moon
26	04:47	First Quarter Moon
27	21:07	Pleiades 1.1°S of the Moon
29	21:53	Moon at perigee = 365,878 km
31	02:31	Jupiter (mag. -2.6) 3.8°S of the Moon
31	13:45	Pollux 3.0°N of the Moon

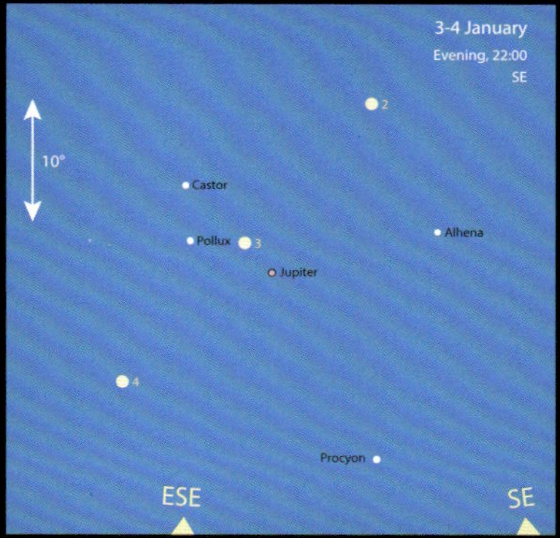

3–4 January • *The Full Moon lies in Gemini close to Jupiter (mag. -2.7) and Pollux, Castor can be seen above the Moon.*

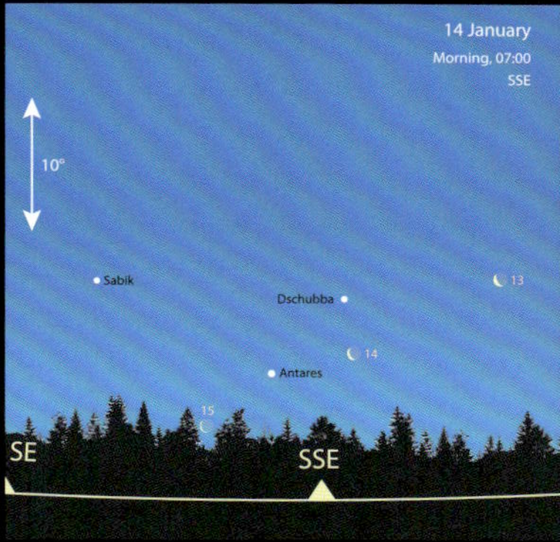

14 January • *Red-orange Antares sits next to the waning crescent Moon in the dawn sky, low in the southeast.*

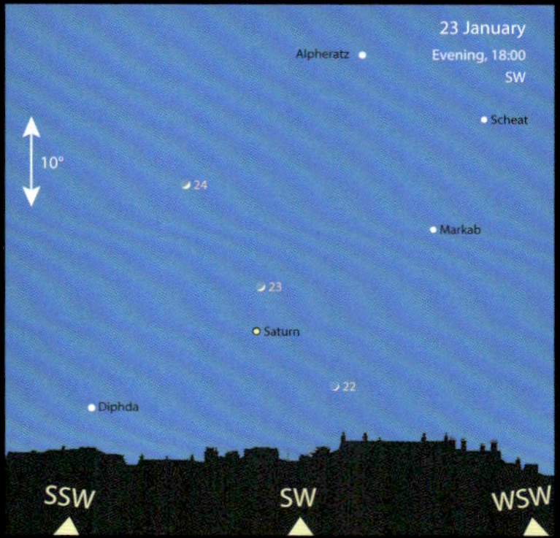

23 January • *The waxing crescent Moon lies above Saturn. The stars of Diphda (β Cet) and Markab (α Peg) are close by.*

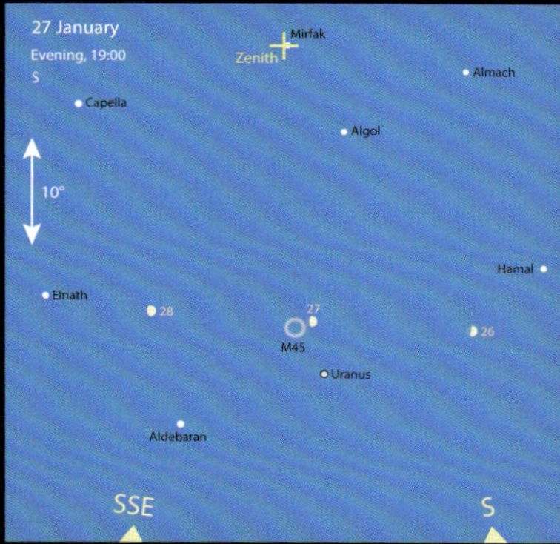

27 January • *The waxing gibbous Moon is next to the Pleiades and Uranus, with Aldebaran, Capella and Elnath to the east.*

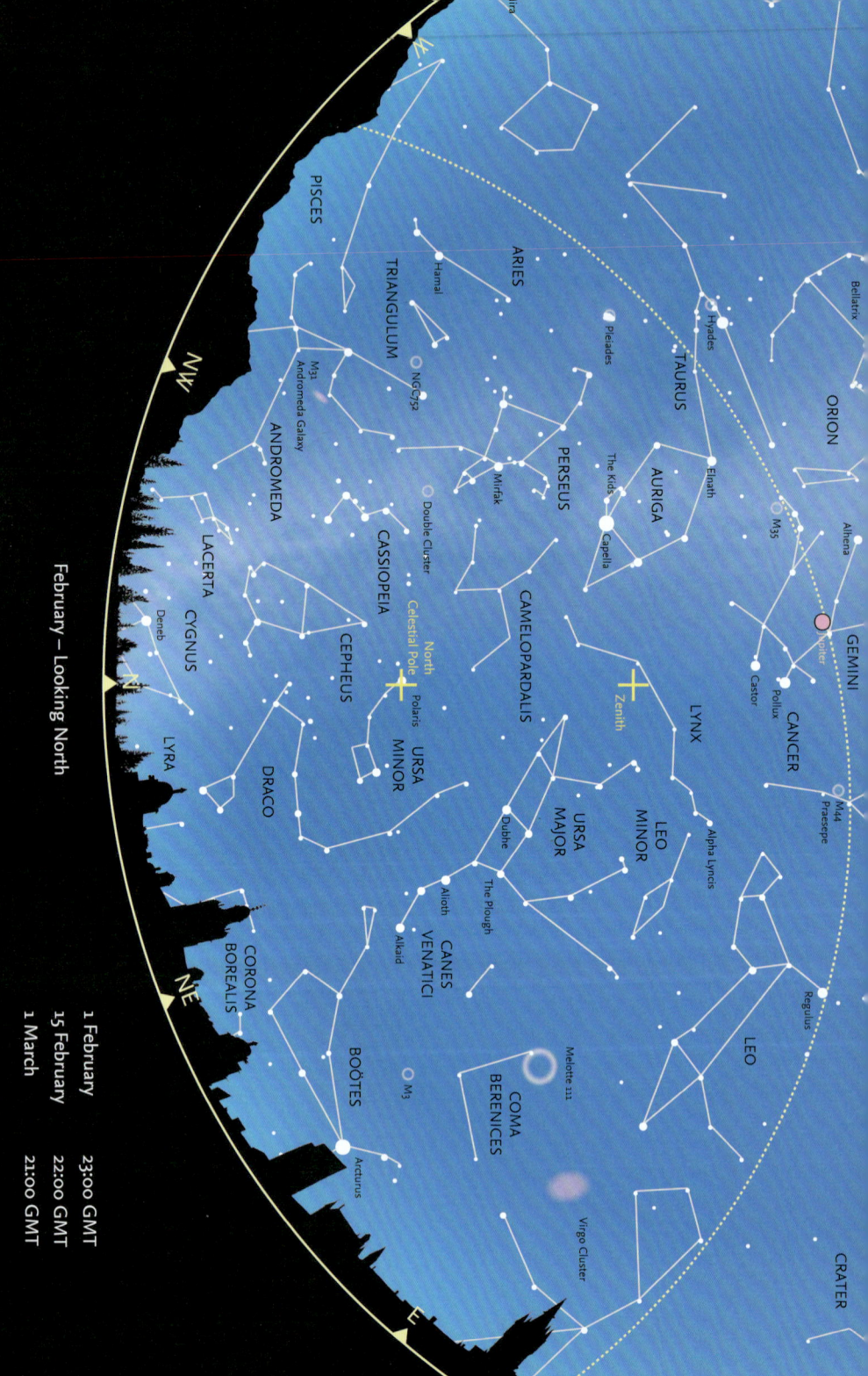

February – Looking North

The denser band of stars of the Milky Way, first observed through a telescope by Galileo Galilei in 1610, run through the northern and western sky. The band stretches from **Cygnus**, low on the northern horizon, through **Cassiopeia**, **Perseus** and **Auriga** and reaching the southern sky via **Gemini** and **Monoceros** (see chart on page 42). Although not as readily visible as the denser star clouds of the summer Milky Way, on a clear night so many stars may be seen that even a distinctive constellation such as Cassiopeia is not immediately obvious.

The head of **Draco** is now higher in the sky and easier to recognize. The blue supergiant **Deneb** (α Cygni), the brightest star in **Cygnus**, may just be visible, almost due north at midnight. Later in the night the constellation of **Boötes** the herdsman, resembling a kite or diamond, clears the horizon. The orange-tinted red giant **Arcturus** (α Boötis) marks an apex. At magnitude −0.05, it is the brightest star in the northern hemisphere of the sky. **Coma Berenices** (Berenice's Hair) is now well above the horizon in the east. Within the constellation is a faint open star cluster of around 40 stars called **Melotte 111** or the Coma Star Cluster (not to be confused with the important Coma Cluster of galaxies, Abell 1656, mentioned on page 55).

On the other side of the sky, in the northwest, most of the constellation of **Andromeda** is still easily seen, although **Alpheratz** (α Andromedae), the star that forms the northeastern corner of the Great Square of **Pegasus** – even though it is actually part of Andromeda – is approaching the horizon and becoming more difficult to detect. High overhead, at the zenith, try to make out the very faint constellation of **Lynx**. It was introduced in 1687 by the famous astronomer Johannes Hevelius to fill the largely blank area between **Auriga**, **Gemini** and **Ursa Major**, and is reputed to be so named because one needed the eyes of a lynx to detect it. The brightest star, α Lyncis has a magnitude of 3.1.

The stars of Cassiopeia in the northern sky.

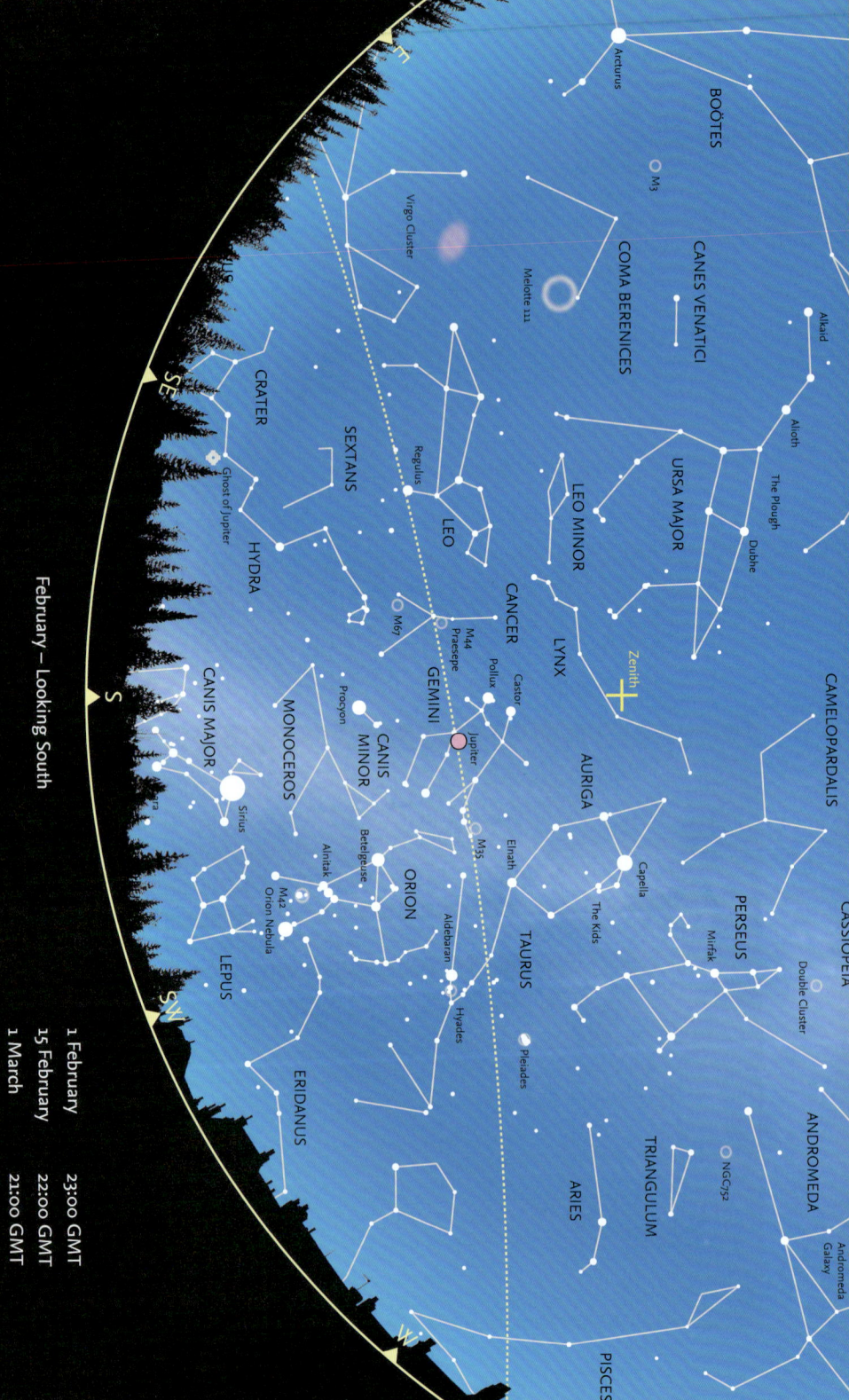

February – Looking South

Apart from **Orion**, the most prominent constellation visible this month is **Gemini**; the heads of the twins are marked by the bright stars **Castor** and **Pollux**, and the lines marking the bodies run southwest towards Orion. Castor (α Geminorum), the fainter star (mag. 1.9), is closer to the North Celestial Pole. Pollux (β Geminorum) is the brighter of the two (mag. 1.2) and it is farther away from the Pole. Pollux lies in the ecliptic and is occasionally occulted by the Moon. Castor is remarkable because it is actually a multiple system, consisting of no fewer than six individual stars arranged in three binary (paired) systems.

Orion's Belt points down to the southeast towards **Sirius**, the brightest star in the sky (at mag. -1.4) in the constellation of **Canis Major**, the whole of which is now clear of the southern horizon. Forming an equilateral triangle with **Betelgeuse** in Orion and Sirius in Canis Major is **Procyon**, the brightest star in the small constellation of **Canis Minor**. Between Canis Major and Canis Minor is the faint constellation of **Monoceros**, which straddles the Milky Way. Although faint, there are many clusters in this area. Directly east of Procyon is the highly distinctive asterism of six stars that form the 'head' of **Hydra**, the largest of all 88 constellations, and which trails such a long way across the sky that it is only around midnight in mid-March that the whole constellation becomes visible.

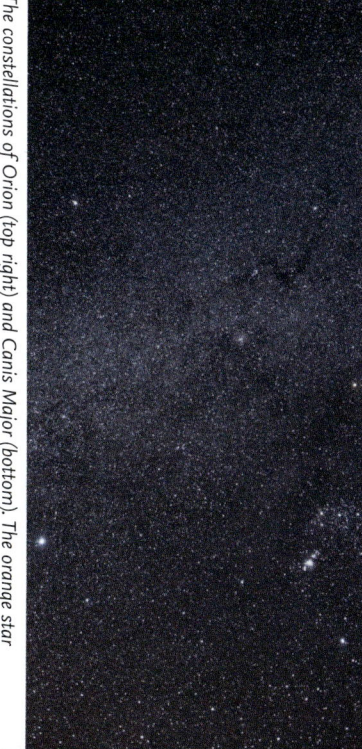

The constellations of Orion (top right) and Canis Major (bottom). The orange star Betelgeuse and the cloudy nebula of Orion can be seen, brilliant Sirius is prominent near the bottom right of the image.

The Moon's phases for February 2026

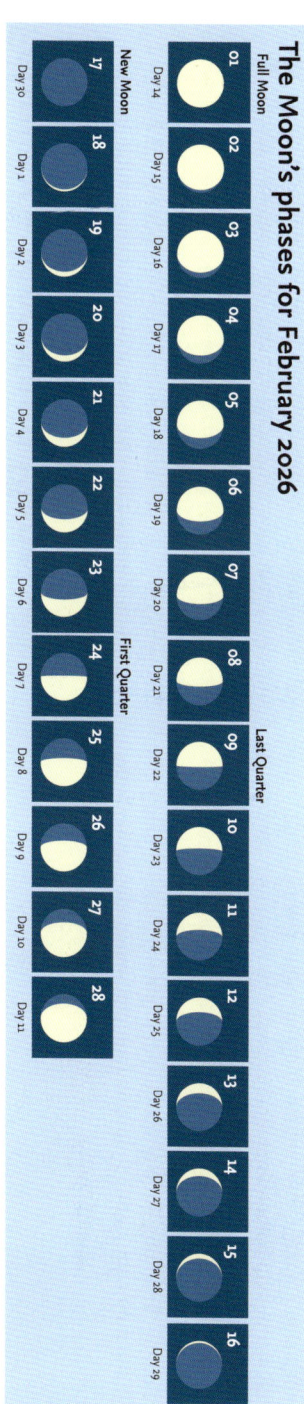

February – Moon and Planets

The Moon

The month begins with a Full Moon, two days later **Regulus** lies 0.4°S of the waning gibbous Moon. On 7 February **Spica** lies 1.8°N of the Moon; the Moon reaches Last Quarter on 9 February. The red supergiant **Antares** can be seen 0.7°N of the waning crescent Moon on 11 February. A New Moon on 17 February is followed by a conjunction with **Mercury** (mag. -0.6) the following day close to the sunset, with the planet only 0.1°N of the thin sliver of waxing crescent Moon. **Saturn** (mag. 1.0) lies 4.6°S of the Moon on 19 February. On 24 February the **Pleiades** lies 1.2°S of the First Quarter Moon. On 27 February **Jupiter** (mag. -2.5) will be placed 4.0°S of the waxing gibbous Moon along with **Pollux** 3.0°N of the Moon.

The Planets

Mercury is in **Capricornus** just after sunset; it moves into **Aquarius** and then **Pisces** in the final week of the month. It dims from mag. -1.2 to 1.5. On 18 February Mercury (mag. -0.6) sits only 0.1°N of the thin waxing crescent Moon, the day after it reaches eastern elongation. **Venus** is also present after sunset, at mag. -3.9 all month. It passes 4.7°S of Mercury on 26 February, a challenging observation due to the glow of the sunset. **Mars** starts in Capricornus with Mercury and moves slowly into Aquarius at the end of the month, at mag 1.1. **Jupiter** stays in **Gemini** and can be seen as soon as skies are dark (mag. -2.6 to -2.4). On 27 February it lies 4°S of the waxing gibbous Moon. **Saturn** is in Pisces and observable for a few hours after sunset. On 16 February it appears 1°S of **Neptune**, at mag. 1.0 with Neptune at mag. 7.8. **Uranus** sits in **Taurus** from sunset until midnight (mag. 5.7) and ends its retrograde motion on 4 February. Neptune lies in Pisces with Saturn (mag. 7.8).

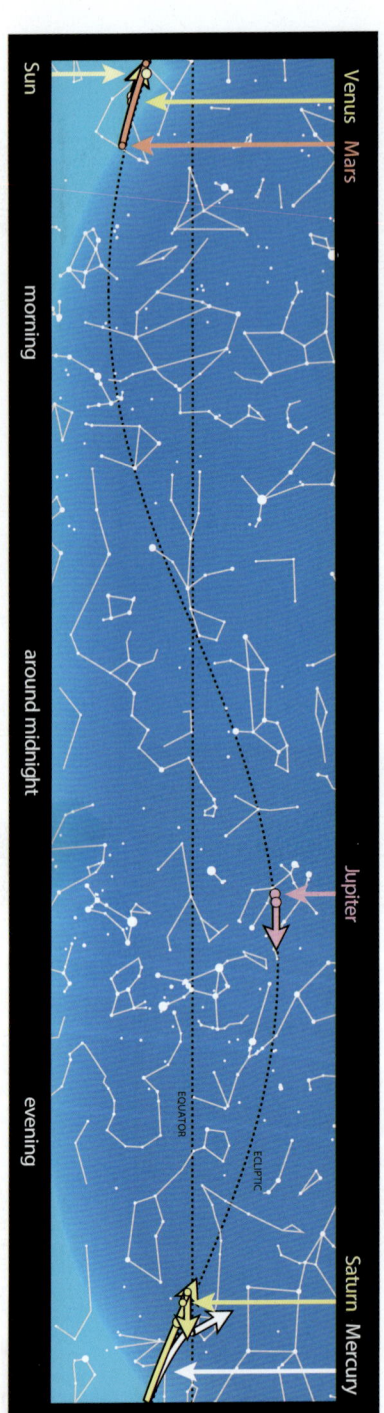

The path of the Sun and the planets along the ecliptic in February.

Calendar for February

01	22:09	Full Moon
03	02:48	Regulus 0.4°S of the Moon
07	08:26	Spica 1.8°N of the Moon
09	12:43	Last Quarter Moon
10	16:52	Moon at apogee = 404,577 km
11	03:19	Antares 0.7°N of the Moon
17	12:01	New Moon
17	12:12	Annular solar eclipse (Antarctica), partial eclipse (south Argentina & Chile, South Africa)
18	23:03	Mercury (mag. −0.6) 0.1°N of the Moon
19	17:39	Mercury at eastern elongation (18.1°E, mag. −0.6)
19	23:54	Saturn (mag. 1.0) 4.6°S of the Moon
24	02:43	Pleiades 1.2°S of the Moon
24	12:28	First Quarter Moon
24	23:18	Moon at perigee = 370,132 km
27	06:26	Jupiter (mag. −2.5) 4.0°S of the Moon
27	21:34	Pollux 3.0°N of the Moon

1–3 February • *The Full Moon on 1 February lies in Cancer and passes through Leo as it wanes; it lies next to Regulus on 3 February.*

11 February • *The waning crescent Moon is next to Antares in Scorpio low in the south at dawn. Sabik is higher up towards the east.*

19 February • *Saturn sits next to the waxing crescent Moon and Mercury (at greatest elongation) at dusk, low in the west.*

27 February • *The waxing gibbous Moon is next to Pollux (in Gemini) and Jupiter. Procyon and Betelgeuse are also visible.*

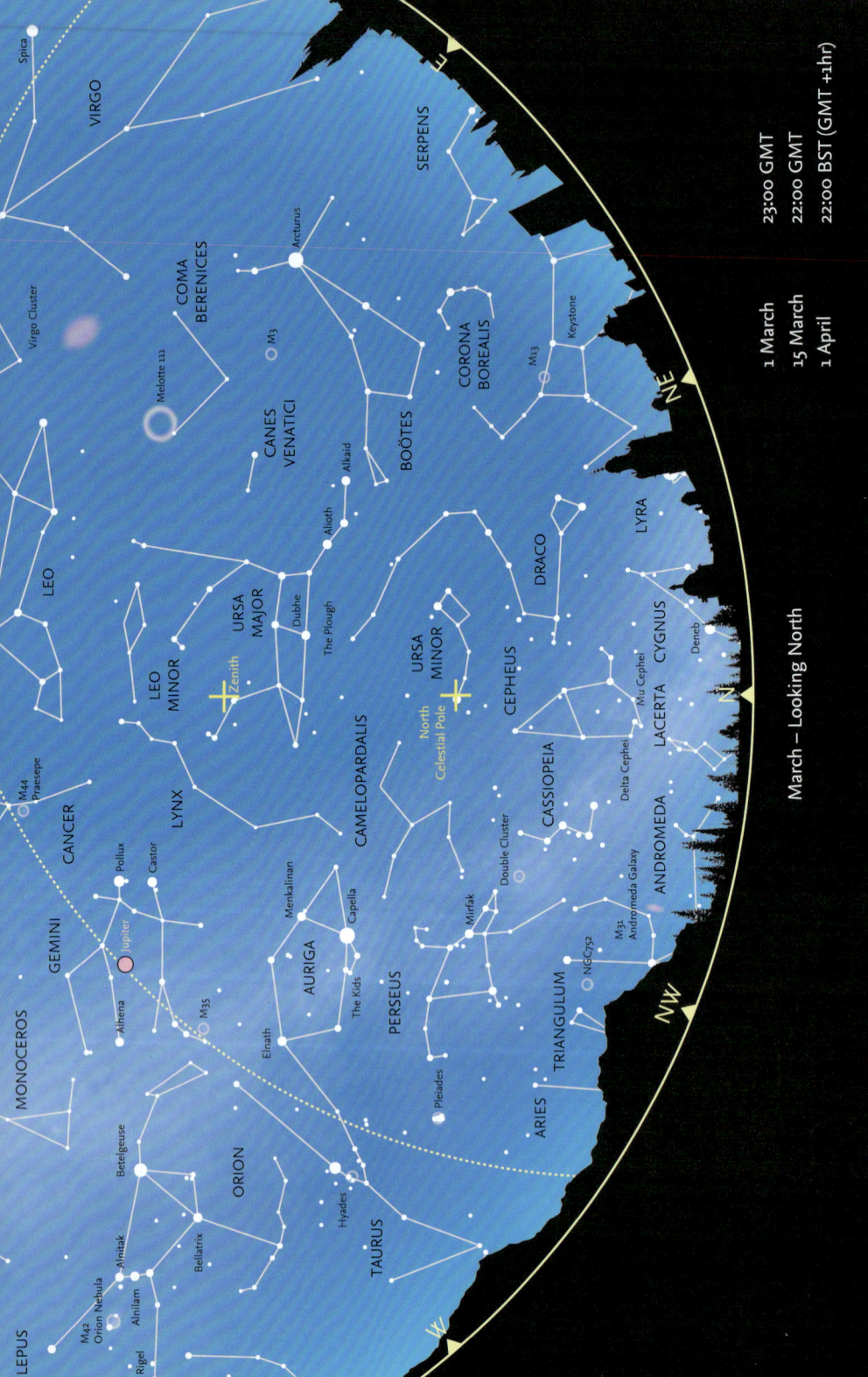

March – Looking North

The Sun crosses the celestial equator on 20 March, at the vernal equinox, when day and night are of almost equal length, and the northern season of spring is considered to begin. The hours of daylight change most rapidly around the equinoxes in March and September. It is also in March that British Summer Time (BST) begins (on Sunday 29 March) so the charts show the appearance at 23:00 for 1 March and 22:00 BST for 1 April. (In the rest of Europe, Daylight Saving Time is introduced on the same date.)

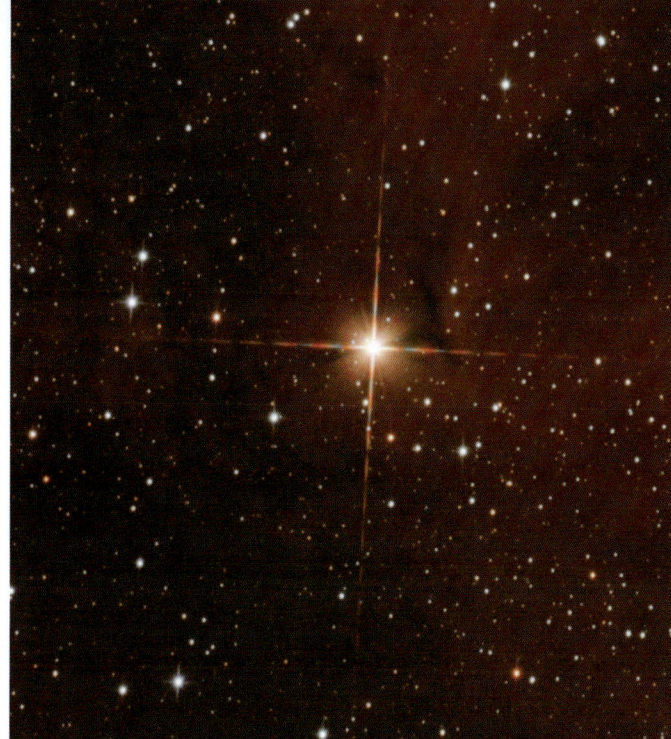

The red supergiant star μ Cephei, also called Herschel's Garnet Star, located in Cepheus.

round towards the northeast, **Vega** (α Lyrae) is marginally higher in the sky. From southern Britain, Deneb is just far enough north to be circumpolar, rising higher in the early hours of the morning.

Early in the month, the constellation of **Cepheus** lies almost due north, with the distinctive 'W' of **Cassiopeia** to its west. Cepheus lies across the border of the Milky Way and is often described as like the gable end of a house or a church tower and steeple. One star stands out because of its deep red colour. This is **μ Cephei**, also known as Herschel's Garnet Star and Erakis. It is a truly gigantic star, a red supergiant with a diameter 1,400 times that of the Sun, and if placed in the Solar System the surface of the star would extend beyond the orbit of Jupiter. **Betelgeuse**, in Orion, is also a red supergiant, however it is one-third the size of μ Cephei.

Another famous and very important star in Cepheus is δ **Cephei**, which is the prototype for the class of variable stars known as Cepheids. These giant stars show a regular variation in their luminosity, and there is a direct relationship between the period of the changes in magnitude and the stars' actual luminosity. By comparing its known luminosity with its apparent brightness, the star's distance can be determined. Once the distances to the first Cepheid variables had been established, examples in more distant galaxies provided information about the scale of the universe. Cepheid variables are the first major 'rung' in the cosmic distance ladder.

Below Cepheus to the east (to the right), it may be possible to catch a glimpse of **Deneb** (α Cygni), just above the horizon. Slightly farther

MARCH 47

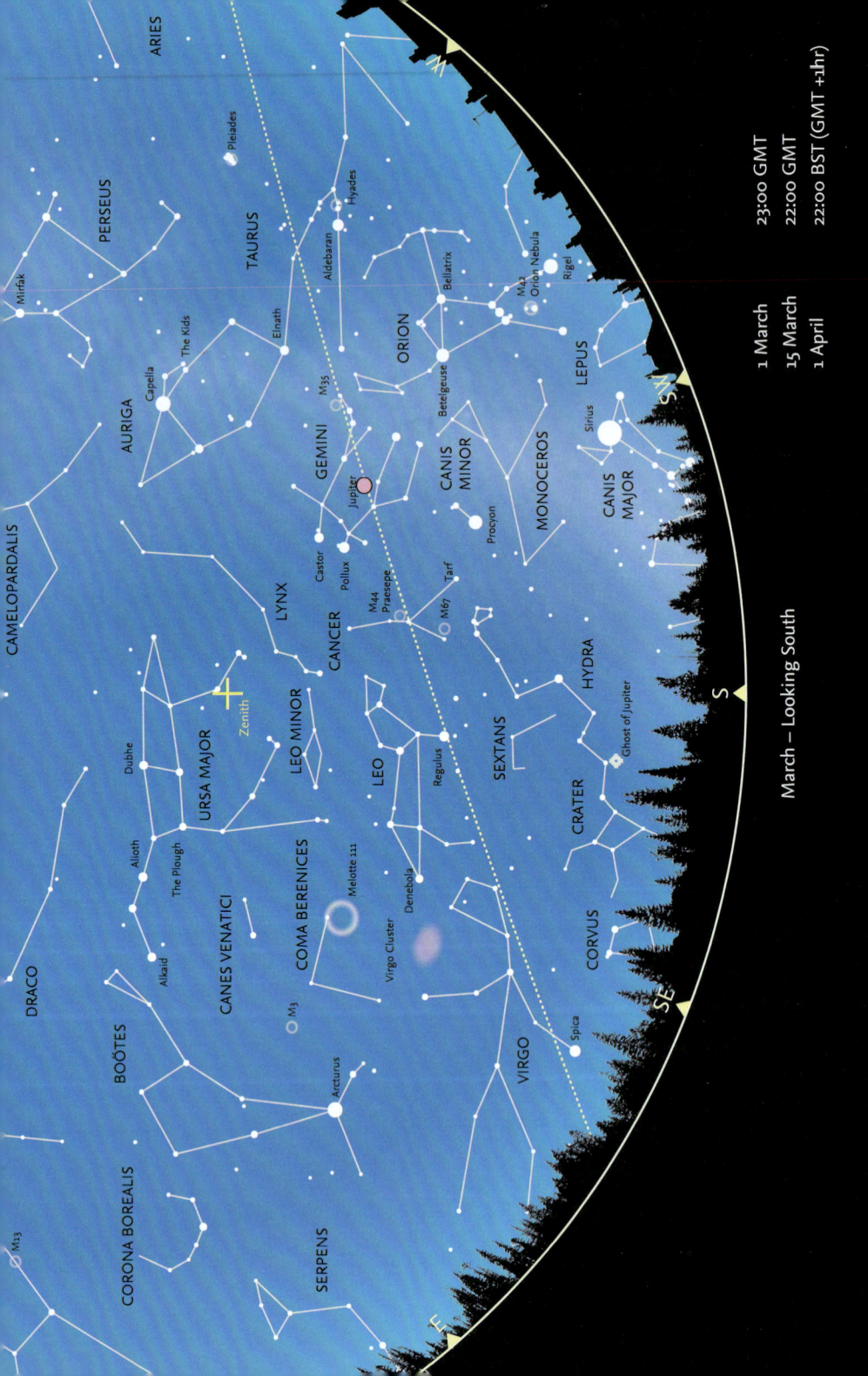

March – Looking South

Lying between the constellations of *Gemini* in the west and *Leo* in the east, and fairly high in the sky above the head of *Hydra*, is the zodiacal constellation of *Cancer*, the crab. Its brightest star is β Cancri (also called *Tarf*), an orange giant around 50 times the size of the Sun. Near the centre of the constellation lies an open cluster, M44 or *Praesepe* (the Manger but also known as the Beehive). On a clear night this cluster of around 1,000 stars, known since antiquity, is just a hazy spot to the naked eye, but appears in binoculars as a group of dozens of individual stars.

Also prominent in March is the constellation of Leo, with the 'backwards question mark' (or the Sickle) of bright stars forming the head of the mythological lion. *Regulus* (α Leonis) – the 'dot' of the question mark or the 'handle' of the sickle and the brightest star in Leo – lies very close to the ecliptic and is one of the few first-magnitude stars that may be occulted by the Moon. *Denebola* (β Leonis), a variable star, marks the tail of the lion.

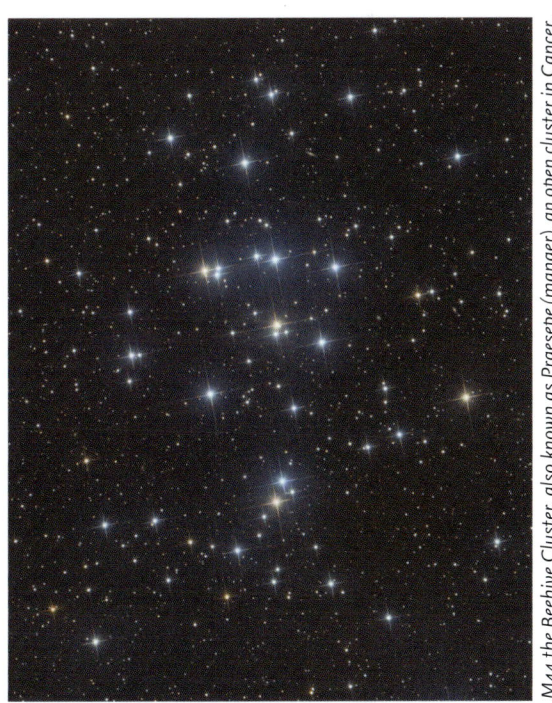

M44 the Beehive Cluster, also known as Praesepe (manger), an open cluster in Cancer.

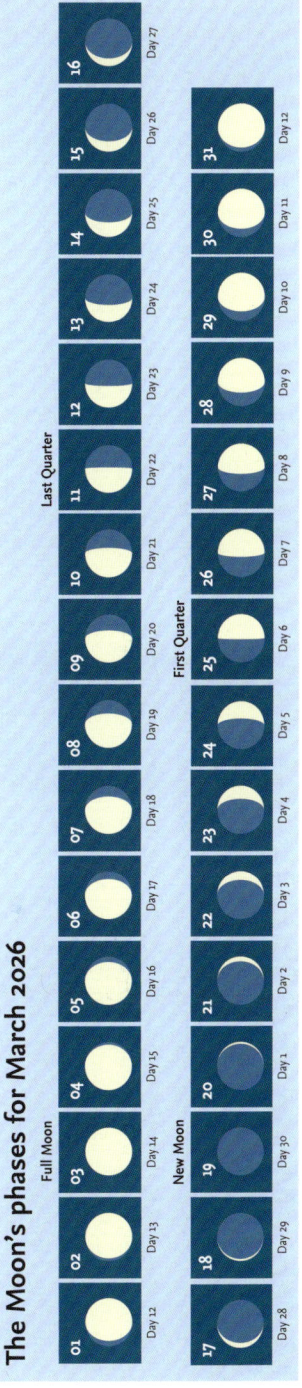

The Moon's phases for March 2026

MARCH 49

March – Moon and Planets

The Moon

On 2 March the waxing gibbous Moon is in *Leo*, 0.4°N of *Regulus*. A Full Moon occurs on 3 March. *Spica* lies 1.8°N of the waning gibbous Moon three days later; on 10 March *Antares* can be seen 0.7°N of the Moon. A day later the Moon is Last Quarter. On 17 March *Mercury* (mag. 1.7) is 2.0°N of the waning crescent Moon and *Mars* (mag. 1.2) is 1.5°S of the Moon. The New Moon appears on 19 March; a day later *Venus* (mag. -3.9) lies 4.6°S of the very thin crescent Moon. On 23 March, two days before it reaches First Quarter, the Moon sits 1.1°N of the *Pleiades*. On 26 March *Jupiter* (mag. -2.2) passes 3.9°S of the waxing gibbous Moon; the following day *Pollux* appears 3.0°N of the Moon. The bright Moon returns to *Regulus* on 29 March, lying 0.4°N of the brightest star in *Leo*.

The Planets

Mercury lies in *Pisces* trailing the Sun at the beginning of the month. After the first week it leads the Sun, eventually appearing early in the dawn sky in *Aquarius*, dimming from mag. 2.0 then brightening to 0.4. On 15 March it joins *Mars* (mag. 1.2) lying 3.4°N. *Venus* enters *Pisces*, dipping into *Cetus* and back as the month progresses and then reaching *Aries* at mag. -3.9. It passes 1°N of *Saturn* (mag. 0.9) on 8 March, visible just after sunset. *Mars* lies in Aquarius for the duration of the month (mag. 1.1 to 1.2). *Jupiter* continues to move slowly through *Gemini* (mag. -2.4 to -2.2). On 11 March it ends retrograde motion and moves eastward. Saturn is in Pisces, setting shortly after sunset and swinging into the dawn sky after 25 March at mag. 0.9. *Uranus* continues in *Taurus*, dimming from mag. 5.7 to 5.8 and *Neptune* in Pisces approaches the Sun and from 22 March moves into the dawn sky (mag. 7.8 to 7.9).

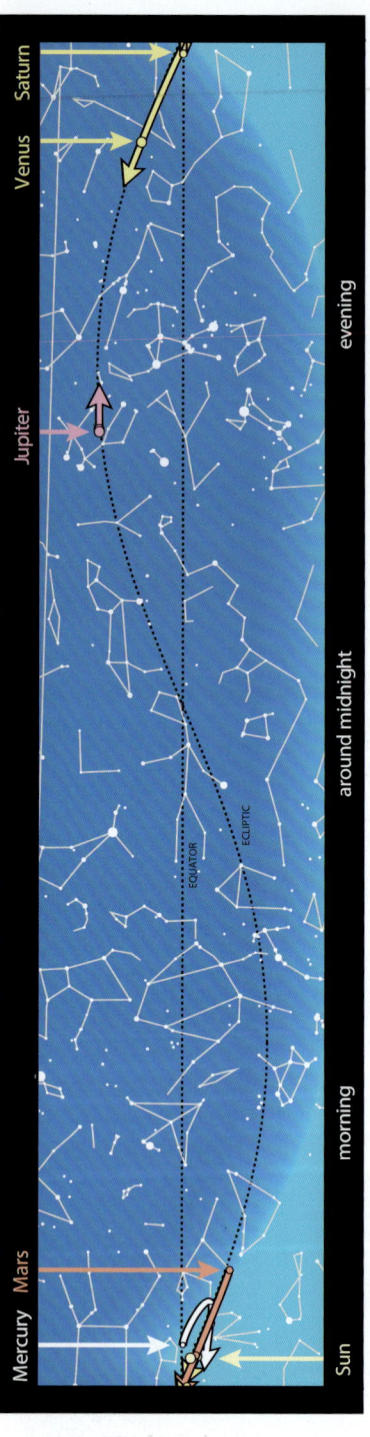

The path of the Sun and the planets along the ecliptic in March.

Calendar for March

02	12:00	Regulus 0.4°S of the Moon
03	11:34	Total lunar eclipse, visible from eastern Asia, Australia, parts of North & South America
03	11:38	Full Moon
06	17:24	Spica 1.8°N of the Moon
10	11:32	Antares 0.7°N of the Moon
10	13:43	Moon at apogee = 404,385 km
11	09:39	Last Quarter Moon
15	19:34	Mercury (mag. 2.7) 3.4°N of Mars (mag. 1.2)
17	14:07	Mercury (mag. 1.7) 2.0°N of the Moon
17	21:51	Mars (mag. 1.2) 1.5°S of the Moon
19	01:23	New Moon
20	12:39	Venus (mag. -3.9) 4.6°S of the Moon
20	14:46	Vernal Equinox
22	11:40	Moon at perigee = 366,858 km
23	08:32	Pleiades 1.1°S of the Moon
25	19:18	First Quarter Moon
26	12:13	Jupiter (mag. -2.2) 3.9°S of the Moon
27	03:18	Pollux 3.0°N of the Moon
29	19:00	Regulus 0.4°S of the Moon. An occultation will be visible from Asia, Africa, Europe (including London) & western Russia

2–3 March • *The Moon passes close to Regulus on 2 March and becomes Full a day later.*

20 March • *Venus lies south of the thin waxing crescent Moon in Pisces.*

25 March • *The First Quarter Moon close to Jupiter in Gemini. Procyon, Pollux, Castor, Capella, Betelgeuse and Aldebaran are close by.*

27 March • *Pollux and the waxing gibbous Moon together with Jupiter in Gemini. Capella and Procyon appear either side.*

MARCH

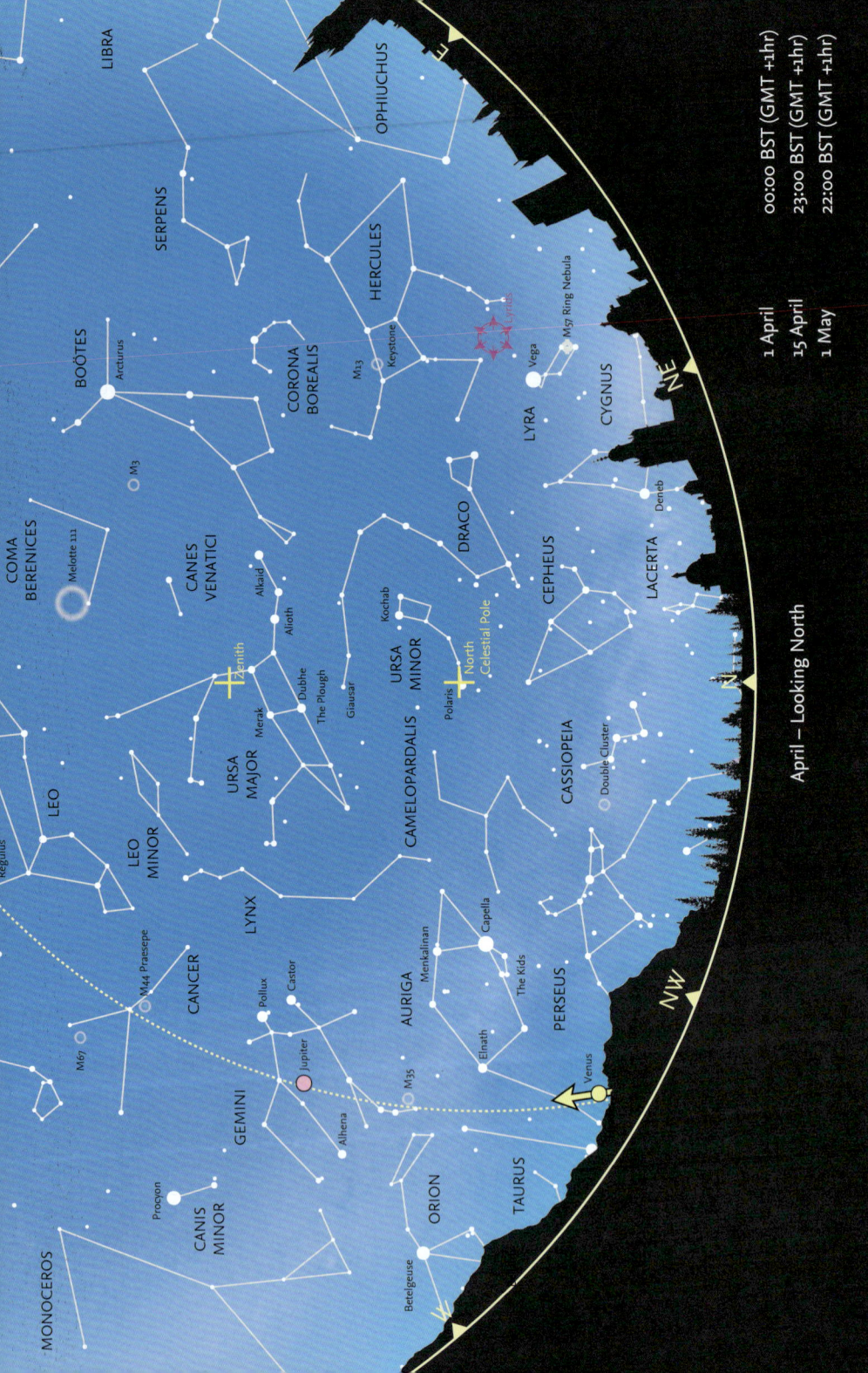

April – Looking North

Cygnus and the brighter regions of the Milky Way are now becoming visible, running more or less parallel with the horizon in the early part of the night. Rising in the northeast is the small constellation of **Lyra** and the distinctive Keystone of **Hercules** above it. This asterism is very useful for locating the globular cluster **M13**, which lies on one side of the quadrilateral. Some of the stars are 12 to 13 billion years old. The winding constellation of **Draco** weaves its way from the four stars that mark its 'head', on the border with Hercules, to end at **Giausar** (λ Draconis) between **Polaris** (α Ursae Minoris) and the Pointers, **Dubhe** and **Merak** (α and β Ursae Majoris, respectively). **Ursa Major** is 'upside down' high overhead, near the zenith. The constellation of **Gemini** stands almost vertically in the west. **Auriga** is still clearly seen in the northwest with its prominent star **Capella**, but, by the end of the month, the southern portion of **Perseus** is starting to dip below the northern horizon. The very faint constellation of **Camelopardalis** lies in the northwest between Polaris and the constellations of Auriga and Perseus.

Meteors

There is one moderate meteor shower this month, the **Lyrids** (often known as the **April Lyrids**). This year the shower begins on 16 April and peaks on 22 April, a few days before the First Quarter Moon. The best time to look for the meteors will be after the Moon has set from 02:00 onwards. At this time the Earth will have turned towards the cloud of debris, and it will be easier to spot them in the sky. There is another stronger shower, the η-**Aquariids**, which begins on 19 April and peaks on 6 May.

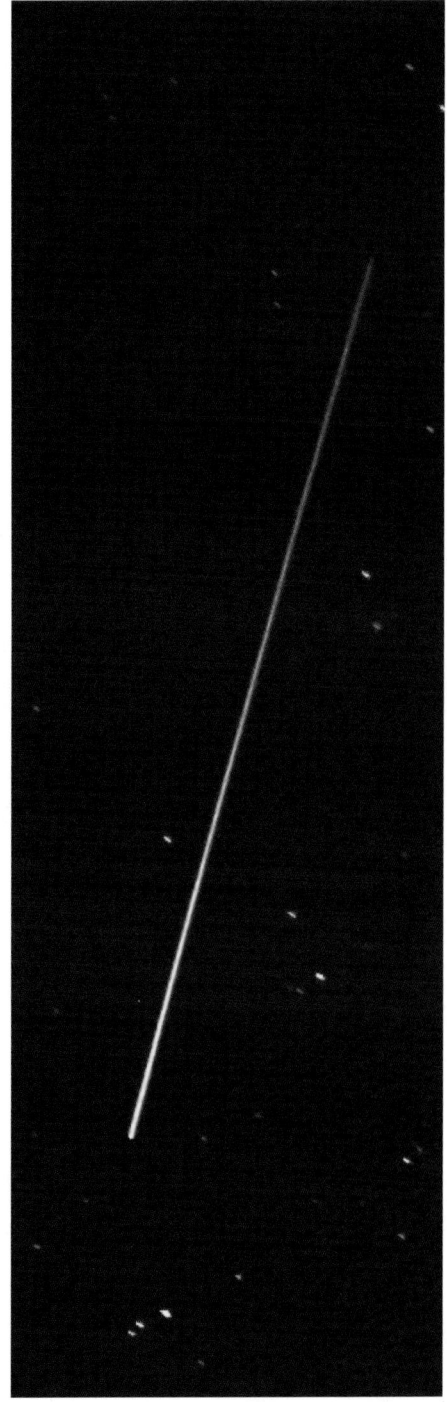

A Lyrid meteor streaks across the sky.

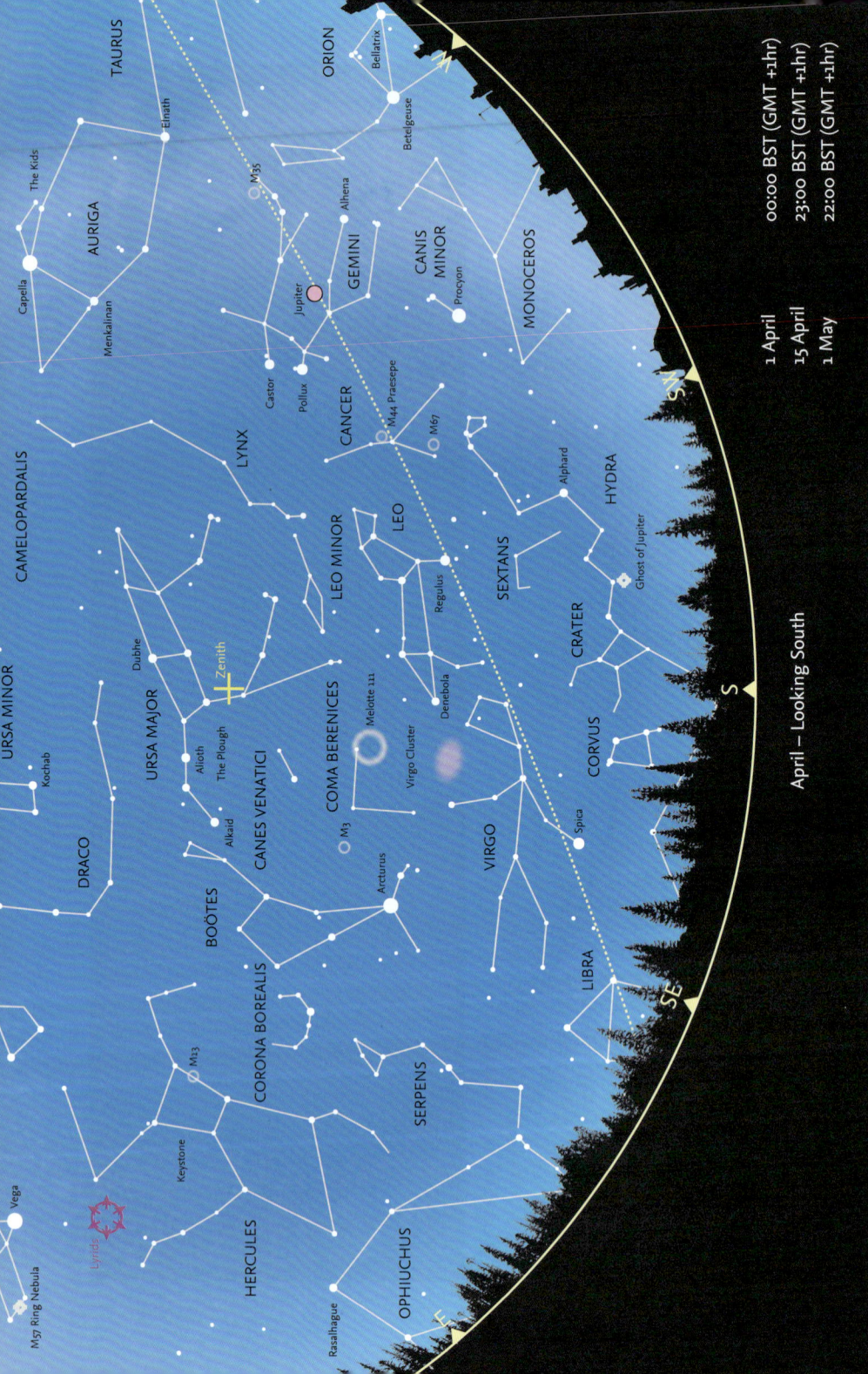

April – Looking South

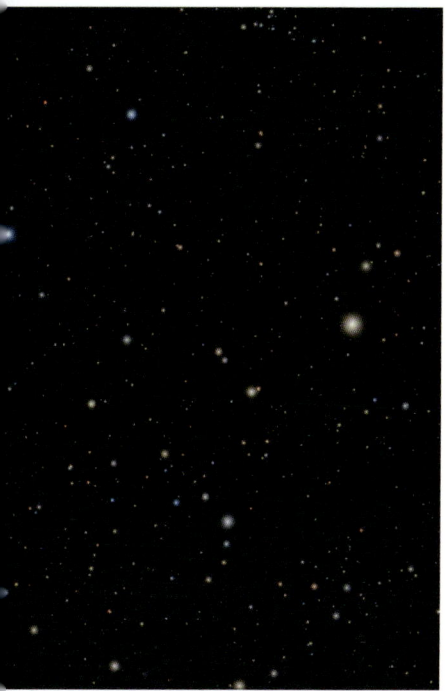

Boötes with the red giant star Arcturus.

Leo is the most prominent constellation in the southern sky in April and looks vaguely like the creature after which it is named. **Gemini**, with **Castor** and **Pollux**, remains clearly visible in the west, and **Cancer** lies between the two constellations. To the east of Leo, the whole of **Virgo**, with **Spica** (α Virginis) its brightest star, is well clear of the horizon. Below Leo and Virgo, the complete length of **Hydra** is visible, running beneath both constellations, with **Alphard** (α Hydrae) halfway between **Regulus** and the southwestern horizon. Farther east, the two small constellations of **Crater** and the rather brighter **Corvus** lie between Hydra and Virgo.

Boötes and **Arcturus** are prominent in the eastern sky, together with the circlet of **Corona Borealis**, framed by Boötes and the neighbouring constellation of **Hercules**. Between Leo and Boötes lies the constellation of **Coma Berenices**, notable for being the location of the open cluster **Melotte 111** (see page 41) and the **Coma Cluster** of galaxies (Abell 1656). There are about 1,000 galaxies in this cluster, which is located near the North Galactic Pole, where we are looking out of the plane of the Galaxy and are thus able to see deep into space. Only about ten of the brightest galaxies in the Coma Cluster are visible with the largest amateur telescopes.

The Moon's phases for April 2026

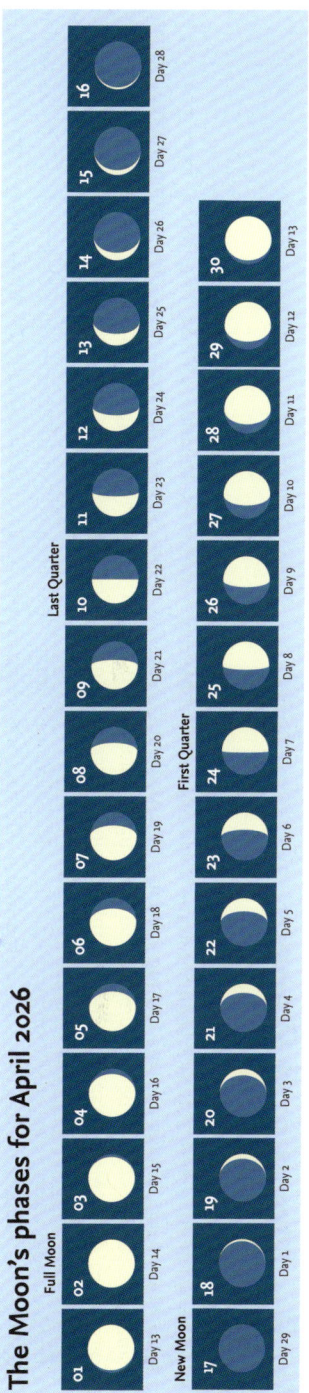

APRIL   55

April – Moon and Planets

The Moon

The Moon is Full on 2 April and as it begins to wane it will pass 1.8°S of **Spica**. On 6 April **Antares** and the waning gibbous Moon will be 0.6° apart, visible a few hours before sunrise. The Last Quarter Moon appears on 10 April. **Mars** (mag. 1.2) and the thin waning crescent Moon are 3.7° apart on 16 April; the following day brings a New Moon. The three-day-old waxing crescent Moon on 19 April lies 4.8°N of **Venus** (mag. -3.9), the **Pleiades** cluster sits 1.0°S of the Moon. On 22 April **Jupiter** (mag. -2.1) can be seen 3.6°S of the Moon; the following day the Moon passes 3.2°S of **Pollux**. On 24 April the Moon is First Quarter. On 26 April the Moon approaches **Regulus**, sitting 0.2°N of the brightest star in **Leo**. The month ends with **Spica** and the waxing gibbous Moon sitting 1.8° apart.

The Planets

Mercury moves from **Aquarius** to **Pisces**, brightening from mag. 0.4 to -0.7. On 3 April Mercury reaches western elongation. Mercury (mag. -0.2) lies 1.7°S of **Mars** (mag. 1.2) and 0.5°S of **Saturn** (mag. 0.9) on 20 April, low on the eastern horizon. **Venus** begins the month in **Aries** and moves into **Taurus** in the latter part of the month. On 8 April Venus (mag. -3.9) sits 4.5°N of (**1**) **Ceres** (mag. 9.0) at dusk and on 24 April it passes within 1°N of **Uranus** (mag. 5.8). Mars moves from Aquarius to Pisces and fluctuates between mag. 1.2 and 1.3. On 19 April it passes 1.2°N of Saturn (mag. 0.9). **Jupiter** stays in **Gemini**, dimming from mag. -2.2 to -2.0. Saturn remains in Pisces and leads the sunrise, pulling away from the Sun as the month progresses (mag. 0.9). Uranus lies in Taurus at mag. 5.8. **Neptune** is in Pisces just before sunrise (mag 7.8).

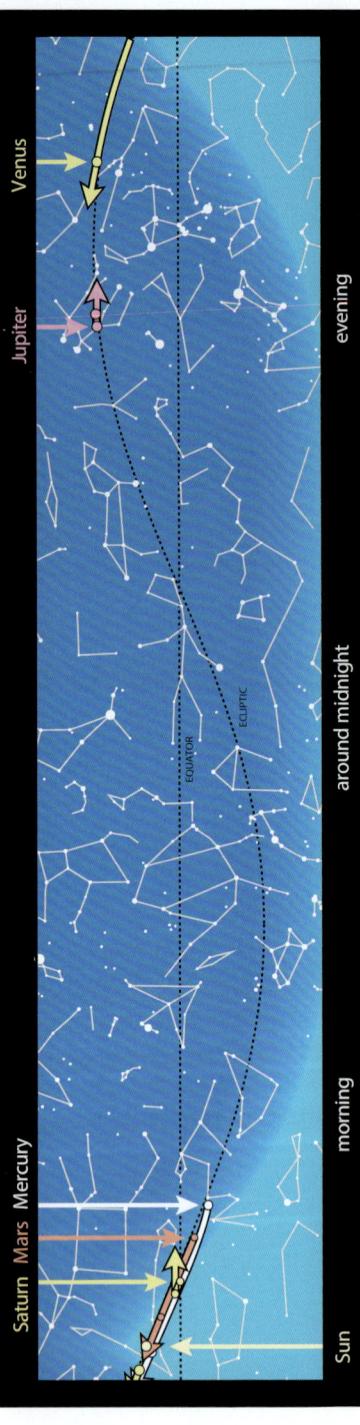

The path of the Sun and the planets along the ecliptic in April.

Calendar for April

02	02:12	Full Moon
03	01:32	Spica 1.8°N of the Moon
03	22:33	Mercury at western elongation (27.8°W, mag. 0.4)
06	19:21	Antares 0.6°N of the Moon
07	08:32	Moon at apogee = 404,974 km
10	04:52	Last Quarter Moon
16	00:45	Mars (mag. 1.2) 3.7°S of the Moon
17	11:52	New Moon
19	06:57	Moon at perigee = 361,631 km
19	08:49	Venus (mag. -3.9) 4.8°S of the Moon
19	16:28	Pleiades 1.0°S of the Moon
19	17:36	Mars (mag. 1.2) 1.2°N of Saturn (mag. 0.8)
20	00:00	Mercury (mag. -0.2) 1.7°S of Mars (mag. 1.2)
20	08:03	Mercury (mag. -0.2) 0.5°S of Saturn (mag. 0.8)
22		Lyrid meteor shower maximum
22	22:06	Jupiter (mag. -2.1) 3.6°S of the Moon
23	08:59	Pollux 3.2°N of the Moon
24	02:32	First Quarter Moon
24	04:17	Venus (mag. -3.9) 3.4°S of the Pleiades
26	00:37	Regulus 0.2°S of the Moon
30	08:17	Spica 1.8°N of the Moon

2–3 April · *The Full Moon passes close to Spica. Arcturus and Denebola are in the same region.*

10 April · *Last Quarter Moon in Sagittarius.*

19 April · *Venus and the waxing crescent Moon in Taurus shortly after sunset.*

21–23 April · *The waxing crescent Moon approaching Gemini and Jupiter. Elnath, Capella and Procyon are nearby.*

APRIL

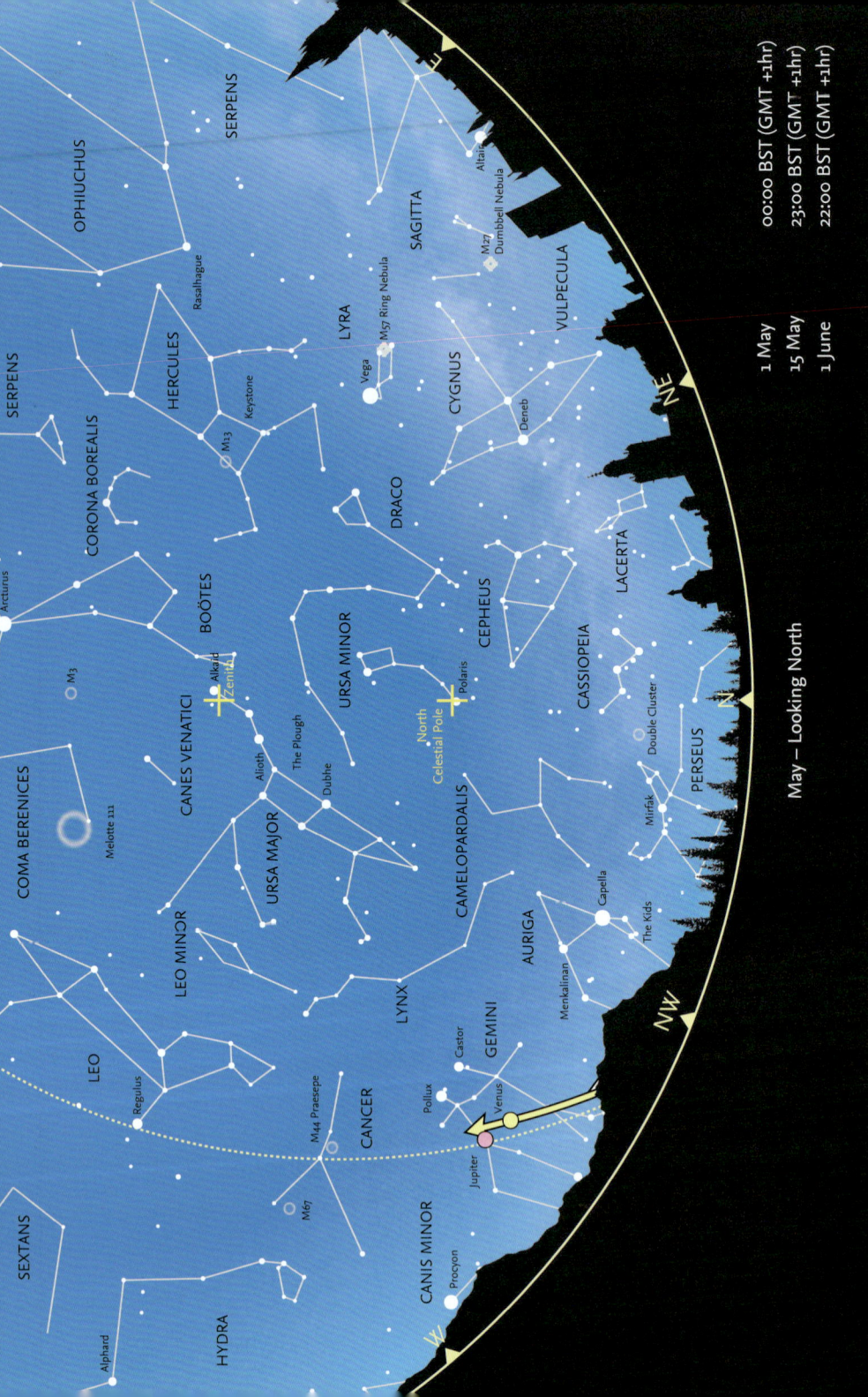

May – Looking North

Cassiopeia is now low over the northern horizon and, to its west, the southern portions of both *Perseus* and *Auriga* are becoming difficult to observe, although the *Double Cluster*, between Perseus and Cassiopeia is still clearly visible. Each cluster is composed of more than 300 supergiants many times hotter than the Sun; some of these stars are nearing the end of their lives and have evolved into red supergiants. The *Andromeda Galaxy* is now too low to be readily visible. The constellations of *Lyra*, *Cepheus*, *Ursa Minor* and the whole of *Draco* are well placed in the sky. *Gemini*, with *Castor* and *Pollux*, is sinking towards the western horizon. *Capella* (α Aurigae) and the three stars in the asterism The Kids are still clear of the horizon.

In the east, two of the stars of the Summer Triangle, *Vega* (α Lyrae) and *Deneb* (α Cygni), are clearly visible, and the third star, *Altair* in *Aquila*, is beginning to climb above the horizon. The whole of *Cygnus* is now visible. The sprawling constellation of *Hercules* is high in the east and the brightest globular cluster in the northern hemisphere, M13, is visible to the naked eye on the western side of the asterism known as the Keystone.

Three faint constellations may be identified before the lighter nights of summer make them difficult objects. Below Cepheus, in the northeastern sky is the zig-zag constellation of *Lacerta*, while to the west, above Perseus and Auriga is *Camelopardalis* and, farther west, the line of faint stars forming *Lynx*.

Later in the night (and in the month) the westernmost stars of *Pegasus* begin to come into view. High overhead, *Alkaid* (η Ursae Majoris), the last star in the 'tail' of the Great Bear, is close to the zenith, while the main body of the constellation has swung round into the western sky.

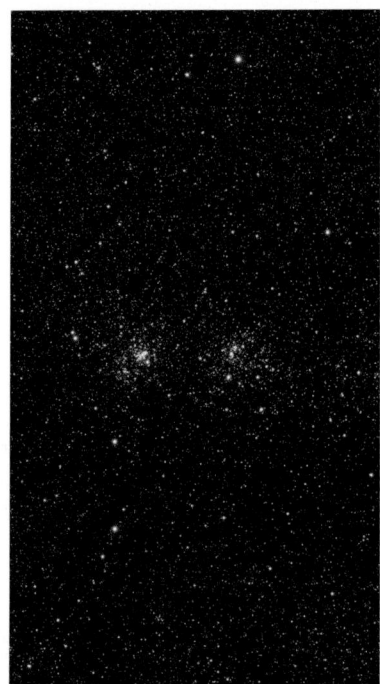

The Double Cluster, composed of the open clusters NGC 869 and NGC 884, in Perseus.

Meteors

The **η-Aquariids** are one of the two meteor showers associated with Comet 1P/Halley, Halley's Comet, last seen in 1986 and not due until 2061. The other shower arising from the cometary debris is the Orionids, in October. The η-Aquariids are not favourably placed for observers in the northern hemisphere, because the radiant is well below the horizon until dawn. However, meteors may still be seen in the eastern sky even when the radiant is below the horizon. There is a radiant map for the η-Aquariids on page 30.

The peak of the shower on 6 May coincides with a bright waning gibbous Moon. The Moon rises just after midnight, observing conditions will be favourable before then, although more meteors are likely to be seen in the early hours of the morning. Maximum hourly rate is about 40 per hour (around half that visible from the northern hemisphere), and a quarter of the meteors leave persistent trains, which can last for several seconds to minutes.

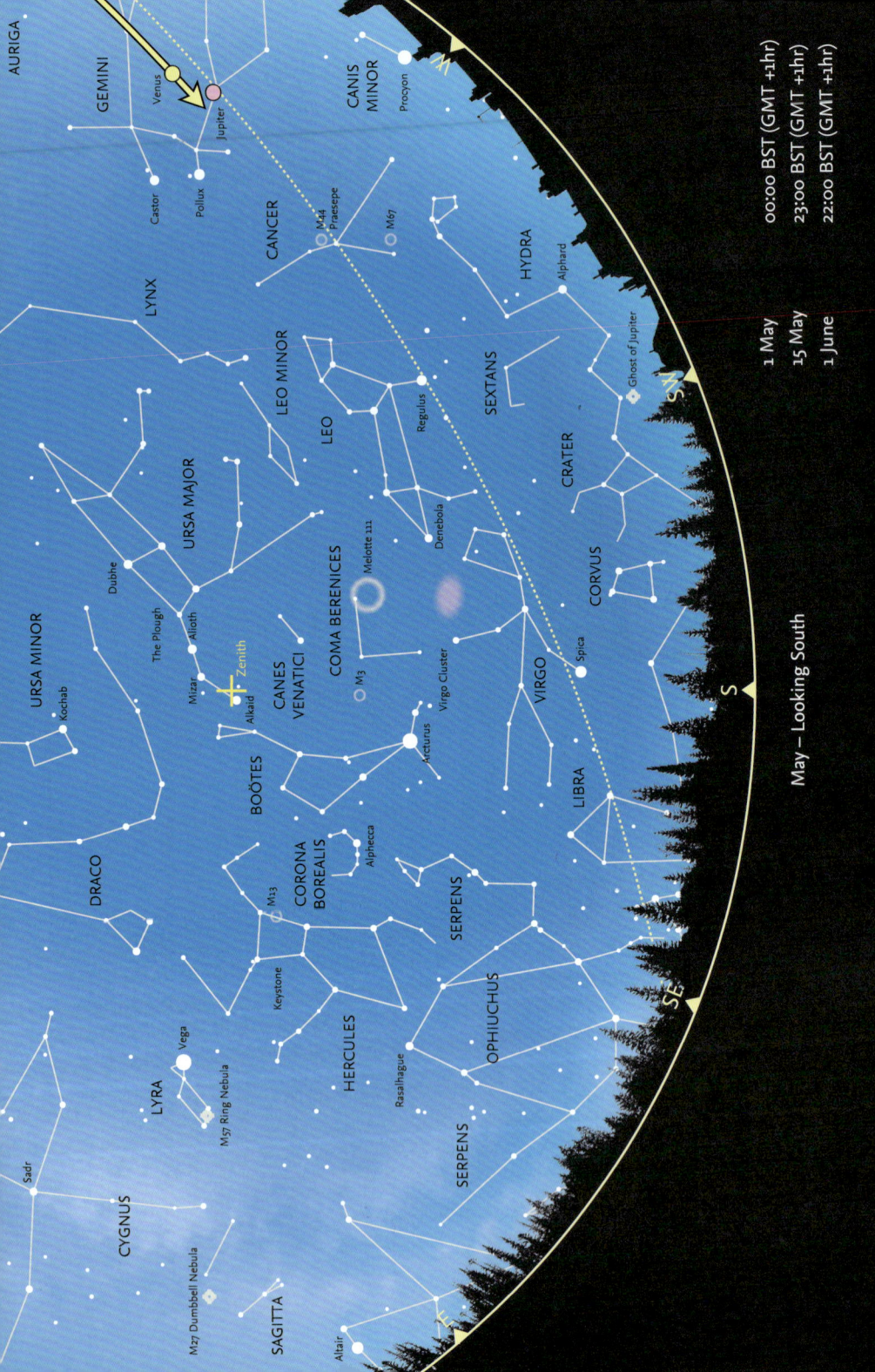

May – Looking South

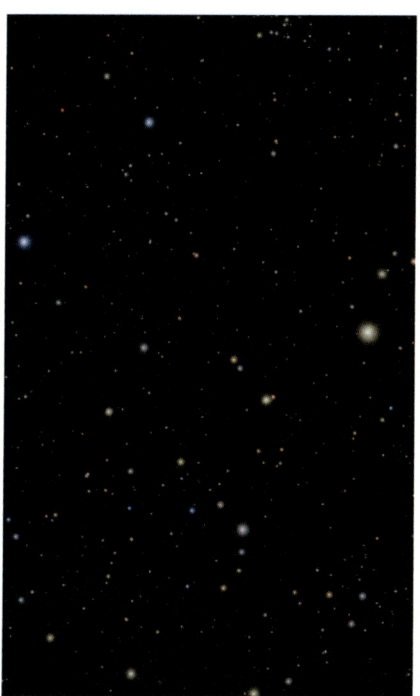

Early in the night, the constellation of *Virgo*, with *Spica* (α Virginis), lies due south, with *Leo* and both *Regulus* and *Denebola* (α and β Leonis, respectively) to its west still well clear of the horizon. Later in the night, the rather faint zodiacal constellation of *Libra* becomes visible and, to its east, the red-orange star *Antares* (α Scorpii) begins to climb up over the horizon.

Virgo contains the nearest large cluster of galaxies, which is the centre of the Local Supercluster, of which the Milky Way galaxy forms part. The Virgo Cluster contains some 2,000 galaxies, the brightest of which are visible in amateur telescopes.

Arcturus in *Boötes* is high in the south, with the distinctive circlet of *Corona Borealis* clearly visible to its east. The brightest star (α Coronae Borealis) is known as *Alphecca*. The large constellation of *Ophiuchus* (which actually crosses the ecliptic and is thus the 'thirteenth' zodiacal constellation) is climbing into the eastern sky. Before the constellation boundaries were formally adopted by the International Astronomical Union in 1930, the southern region of Ophiuchus was regarded as forming part of the constellation of *Scorpius*, which had been part of the zodiac since antiquity.

The constellation of Boötes with the red giant, Arcturus – a star in the latter stages of its lifecycle.

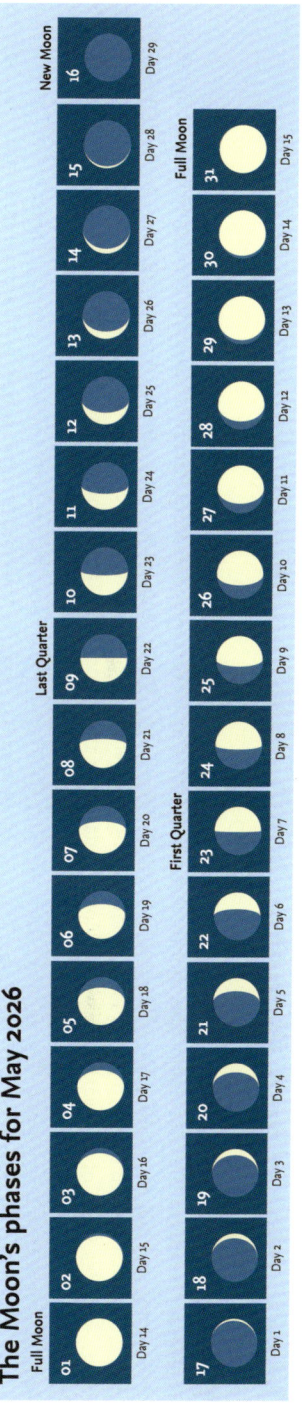

The Moon's phases for May 2026

May – Moon and Planets

The Moon

The Moon is full on the first day of the month. *Antares* is 0.5°N of the waning gibbous Moon on 4 May. The Last Quarter Moon will be visible on 9 May and 16 May brings the New Moon. On 19 May *Venus* (mag. -3.9) lies 2.9°S of the waxing crescent Moon, a day later the Moon swings 3.1°N of *Jupiter* (mag. -1.9) and 3.4°S of *Pollux*. On 23 May *Regulus* is less than a tenth of a degree apart from the First Quarter Moon and on 27 May *Spica* is 1.9°N of the Moon. The end of the month brings a Full Moon with *Antares* 0.4°N.

The Planets

Mercury moves from *Pisces* into *Aries* and eventually *Taurus*, brightening from mag. -0.8 to -2.4 and then dimming to -0.6. In the latter half of the month Mercury appears shortly after sunset. *Venus* starts the month in Taurus and moves into *Gemini*, visible after sunset (mag. -3.9 to -4.0). *Mars* starts in Pisces and passes into Aries, appearing at dawn a few hours before sunrise (mag. 1.2 to 1.3). *Jupiter* lies in Gemini and is above the horizon from sunset until around midnight (mag. -2.0 to -1.9). *Saturn* lies on the border between *Cetus* and Pisces, visible just before sunrise (mag. 0.9). *Uranus* in Taurus approaches the Sun and moves into the dawn sky after 22 May (mag. 5.8). *Neptune* is low in the east before sunrise in Pisces at mag 7.8.

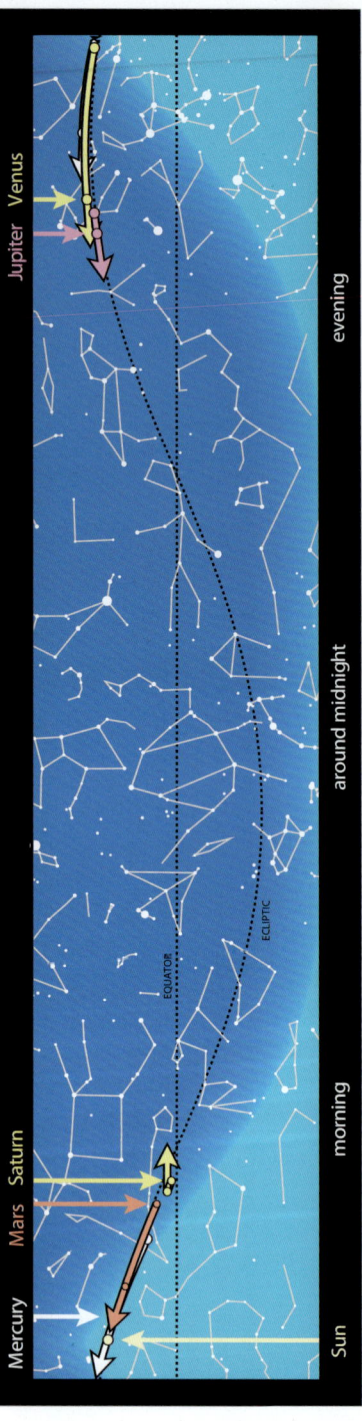

The path of the Sun and the planets along the ecliptic in May.

Calendar for May

01	17:23	Full Moon
04	02:20	Antares 0.5°N of the Moon
04	22:30	Moon at apogee = 405,843 km
06		Eta Aquariid meteor shower maximum
09	21:10	Last Quarter Moon
16	20:01	New Moon
17	13:48	Moon at perigee = 358,074 km
19	01:50	Venus (mag. -3.9) 2.9°S of the Moon
20	12:39	Jupiter (mag. -1.9) 3.1°S of the Moon
20	16:30	Pollux 3.4°N of the Moon
23	06:41	Regulus 0.0°N of the Moon
23	11:11	First Quarter Moon
27	14:09	Spica 1.9°N of the Moon
31	08:32	Antares 0.4°N of the Moon
31	08:45	Full Moon

6 May • *The waning gibbous Moon in Sagittarius in the south, the Eta Aquariids meteor shower in the dawn sky.*

19–21 May • *The waxing crescent Moon in Gemini with Jupiter and Venus, Pollux and Castor. Procyon and Capella lie close by.*

23 May • *The First Quarter Moon close to Regulus and Denebola in Leo.*

30 May • *Saturn rising in the east along with Mars just before sunrise. Alpheratz (α And) sits higher up.*

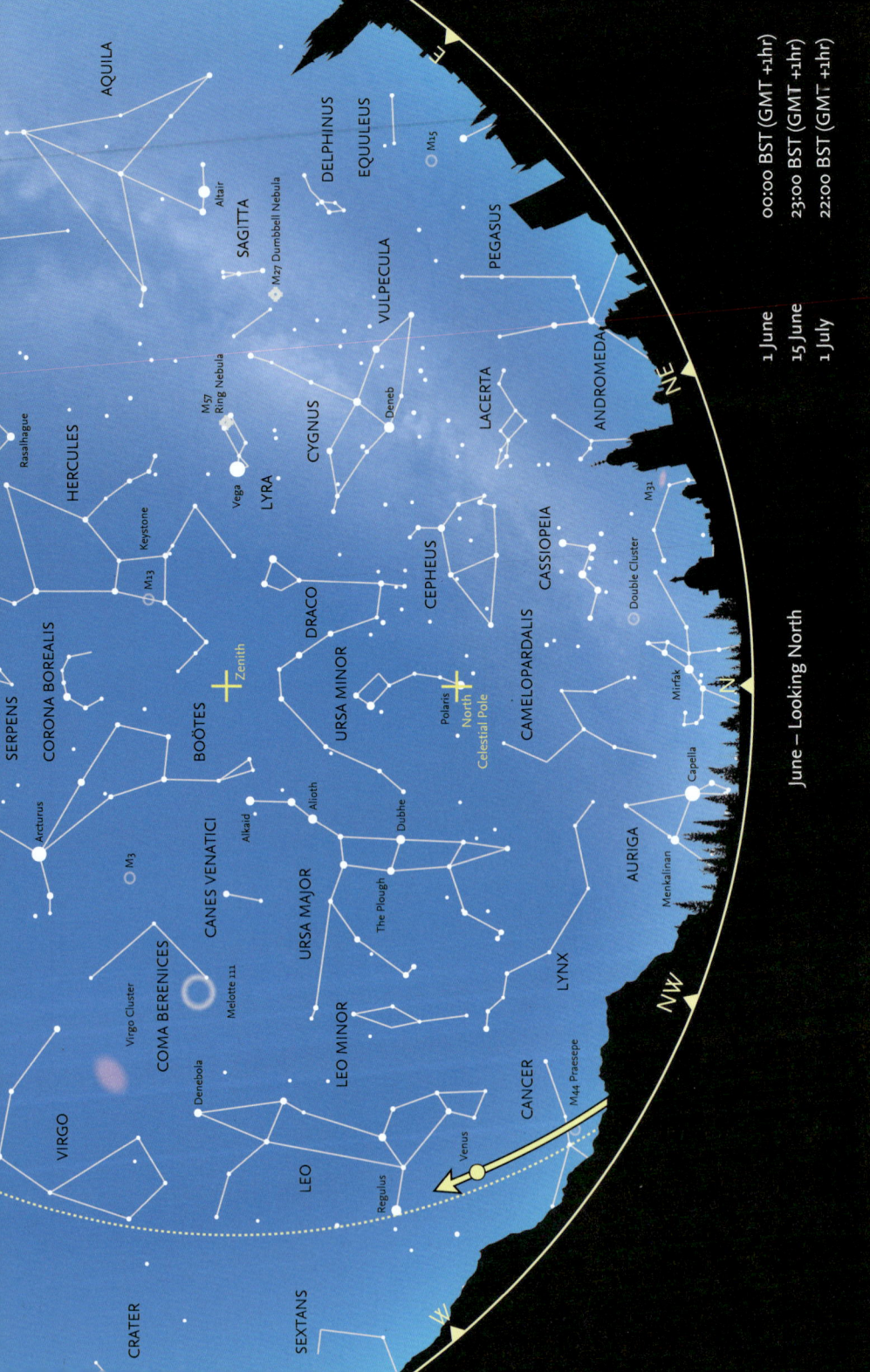

June – Looking North

Around summer solstice (21 June), even in southern England and Ireland, a form of twilight persists throughout the night due to the Sun not setting as far below the horizon. Farther north, in Scotland, the sky remains so light that most of the fainter stars and constellations are invisible. There, even brighter stars such as the seven stars making up the well-known asterism known as the Plough in **Ursa Major** may be difficult to detect except around local midnight, 00:00 UT (01:00 BST).

But there is one compensation during these light nights: even southern observers may be lucky enough to witness a display of noctilucent ('night-shining') clouds. These are highly distinctive clouds shining with an electric-blue tint, observed in the sky in the direction of the North Pole. They are the highest clouds in the atmosphere, occurring at altitudes of 80–85 km, and they are composed of ice crystals. They are only visible during summer nights shortly after sunset and before sunrise, for up to six weeks on either side of the solstice, but the clouds themselves remain illuminated by sunlight. They act like high mirrors, reflecting sun rays when the Sun is below the horizon.

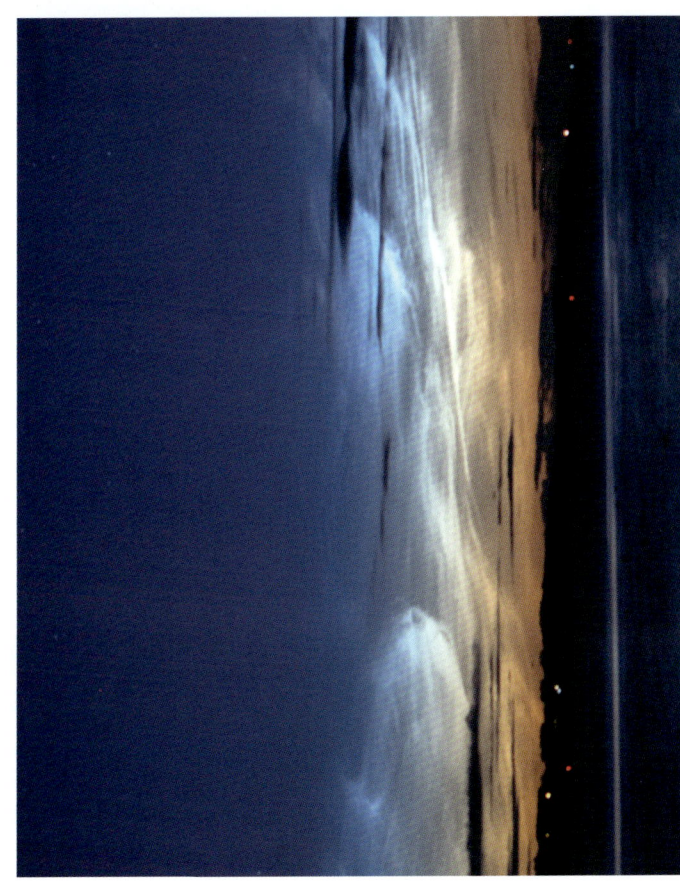

Noctilucent clouds and Comet Neowise over Aberogwen, Wales (2020).

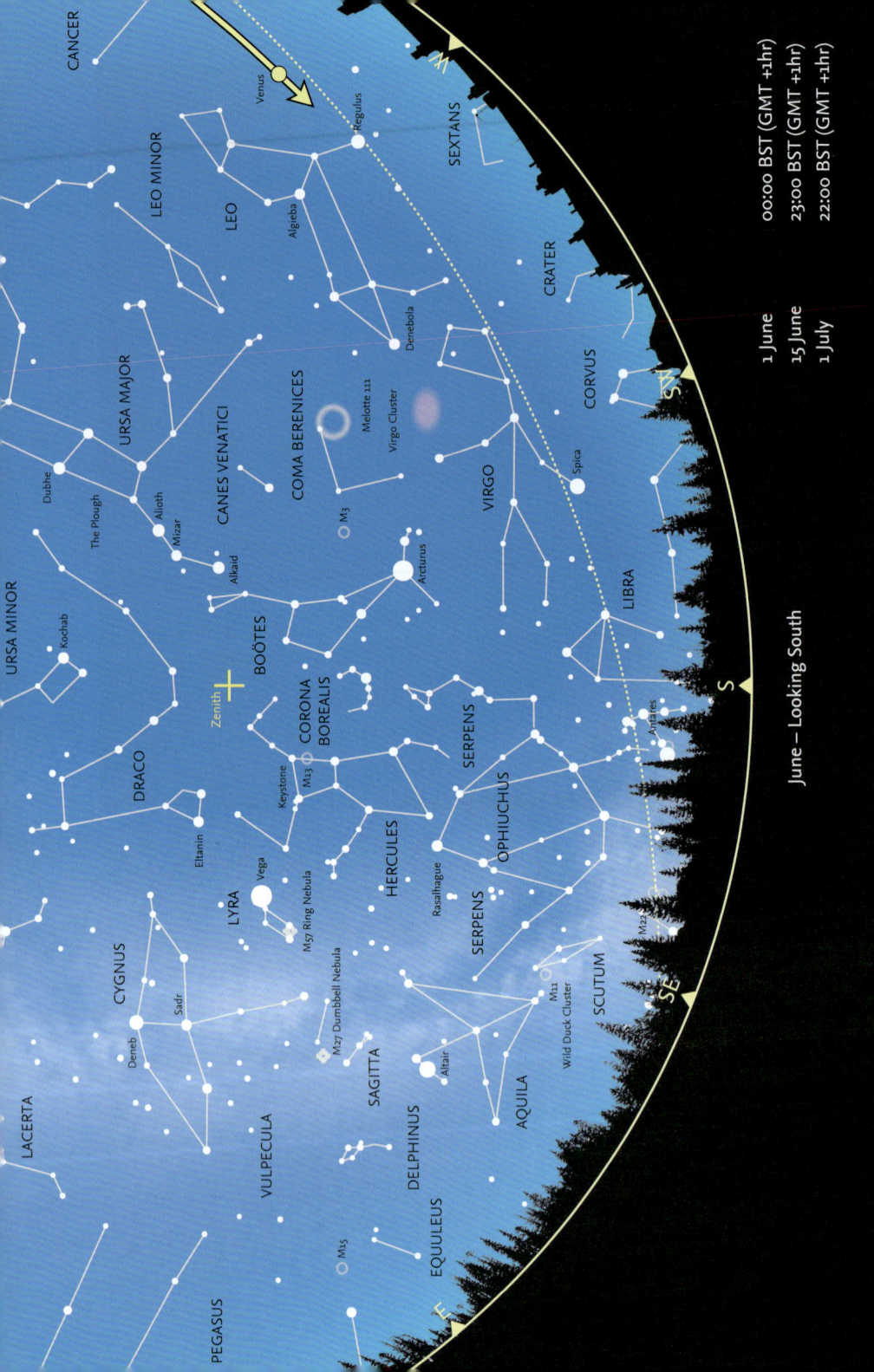

June – Looking South

Although the persistent twilight makes observing even the southern sky difficult, there is still plenty to see after dark. **Deneb**, **Vega** and **Altair** forming the Summer Triangle are the first bright stars to be visible in the east. All three are larger than the Sun, Deneb being 200 times greater in size, and up to 200,000 times more luminous. Lying 100 times further away than the other two stars, we see all three having similar apparent brightness in the summer sky. If Deneb was the same distance as Vega and Altair, it would be so bright in our sky it would cast shadows, like the Full Moon.

The zodiacal constellation of **Libra** lies almost due south. The red supergiant star **Antares** – the name means the 'rival of Mars' – in **Scorpius** is visible slightly to the east of Libra, but the 'tail' or 'sting' remains below the horizon. Higher in the sky is the large constellation of **Ophiuchus** (the Serpent Bearer), lying between the two halves of the constellation of **Serpens**: **Serpens Caput** (Head of the Serpent) to the west and **Serpens Cauda** (Tail of the Serpent) to the east. The ecliptic runs across Ophiuchus, and the Sun actually spends far more time in the constellation than it does in the 'classical' zodiacal constellation of Scorpius, a small area of which lies between Libra and Ophiuchus.

Higher in the southern sky, the three constellations of **Boötes**, **Corona Borealis** and **Hercules** are now better placed for observation than at any other time of the year. This is an ideal time to observe the fine globular cluster of **M13** in Hercules with binoculars or a telescope.

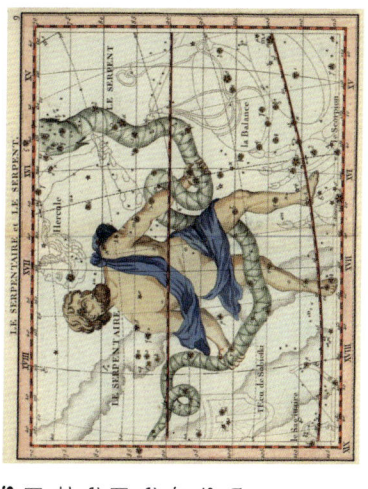

Ophiuchus and Serpens. The ecliptic passes through Ophiuchus.

The Moon's phases for June 2026

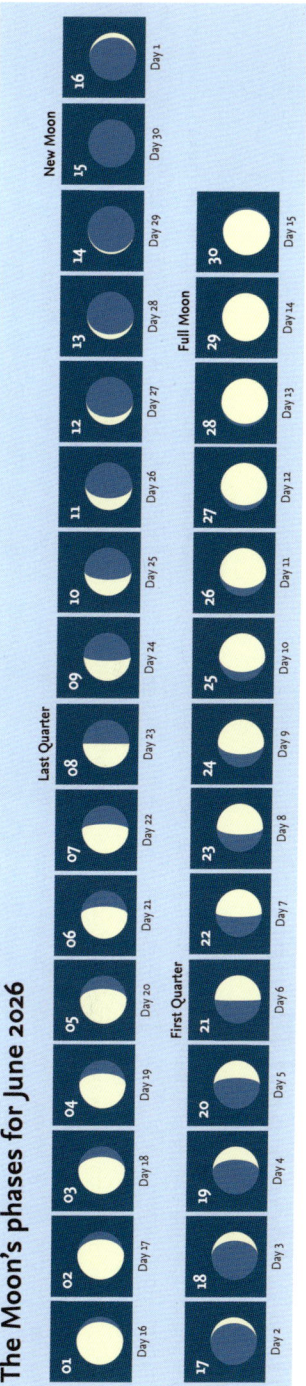

JUNE **67**

June – Moon and Planets

The Moon

The Moon reaches Last Quarter on 8 June. On 13 June the waning crescent Moon passes 1.0°N of the **Pleiades**. A New Moon occurs on 15 June; a day later **Mercury** (mag. 0.5) lies 2.6°S of a thin sliver of the Moon. The following day the waxing crescent Moon is 3.6°S of **Pollux**, 2.5°N of **Jupiter** (mag. -1.8) and 0.3°N of **Venus** (mag. -4.0). On 19 June **Regulus** is 0.3°N of the Moon. The First Quarter Moon is visible on 21 June at the summer solstice. Two days later the waxing gibbous Moon passes 2.2°S of **Spica**. On 27 June **Antares** lies 0.5°N of the Moon and on 29 June the Moon is Full.

The Planets

Mercury is in **Gemini** and appears after sunset as the month progresses (mag. -0.5 to 2.0). On 15 June it is at eastern elongation (mag. 0.5). On 25 June Mercury (mag. 1.3) passes 3.8°S of **Jupiter** (mag. -1.8). **Venus** starts the month in **Gemini**, 12 days later it moves into **Cancer** and by the end of the month it is in **Leo** and visible in the evening (mag. -4.0 to -4.1). On 7 June Venus is 4.6°S of **Pollux** and on 9 June it sits 1.6°N of Jupiter (mag. -1.9). **Mars** is in **Aries** and progresses into **Taurus** in the last week of the month. It's visible up to a few hours before sunrise (mag. 1.3 to 1.4). On 28 June Mars is 4.3°S of the **Pleiades** (mag. 1.4). Jupiter moves through Gemini and into Cancer, visible from sunset until around 22:00 (mag. -1.9 to -1.8). **Saturn** sits in **Pisces**, appearing a few hours after midnight (mag. 0.9 to 0.8). **Uranus** is in **Taurus** and only observable around an hour before sunrise (mag. 5.8). **Neptune** lies in Pisces and above the horizon several hours after midnight (mag. 7.8).

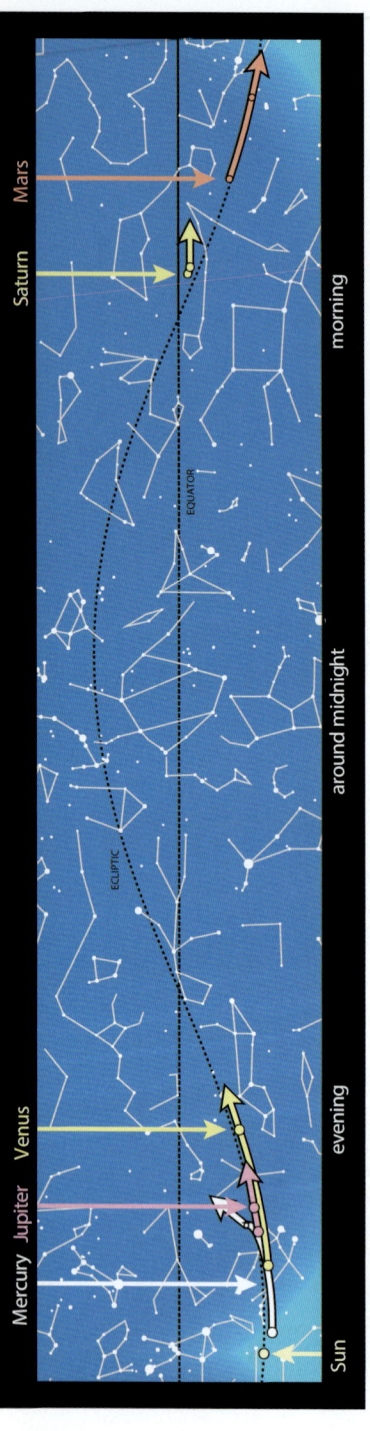

The path of the Sun and the planets along the ecliptic in June.

Calendar for June

01	04:32	Moon at apogee = 406,369 km
07	16:17	Venus (mag. -4.0) 4.6°S of Pollux
08	10:00	Last Quarter Moon
09	20:03	Venus (mag. -4.0) 1.6°N of Jupiter (mag. -1.9)
13	13:15	Pleiades 1.0°S of the Moon
14	23:18	Moon at perigee = 357,196 km
15	02:54	New Moon
15	20:00	Mercury at eastern elongation (24.5°E, mag. 0.5)
16	19:32	Mercury (mag. 0.5) 2.6°S of the Moon
17	02:08	Pollux 3.6°N of the Moon
17	06:54	Jupiter (mag. -1.8) 2.5°S of the Moon
17	20:21	Venus (mag. -4.0) 0.3°S of the Moon
19	14:31	Regulus 0.3°N of the Moon
21	08:25	Summer Solstice
21	21:55	First Quarter Moon
23	20:11	Spica 2.2°N of the Moon
25	12:00	Mercury (mag. 1.3) 3.8°S of Jupiter (mag. -1.8)
27	14:32	Antares 0.5°N of the Moon
28	07:11	Moon at apogee = 406,267 km
28	18:32	Mars (mag. 1.3) 4.3°S of the Pleiades
29	23:57	Full Moon

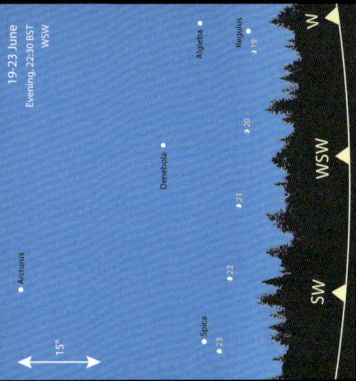

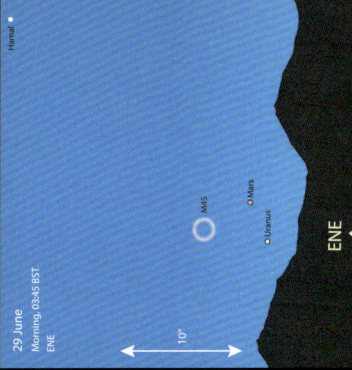

9 June • Venus and Jupiter appear close together in Gemini, with Mercury nearby. Pollux and Castor shine brightly just above.

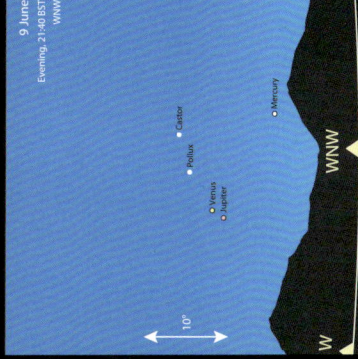

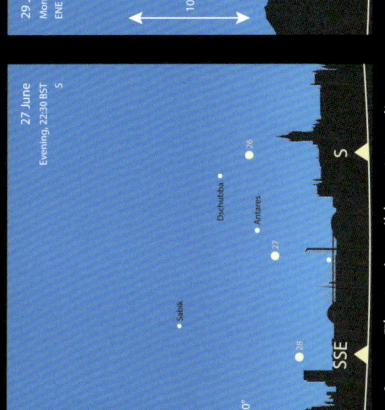

19-23 June • The Moon progresses eastward from Regulus to Spica, reaching First Quarter on 21 June.

27 June • The waxing gibbous Moon lies to the east of Antares, the rival of Mars.

29 June • Mars, Uranus and the Pleiades all in Taurus rising in the east just before sunrise.

JUNE 69

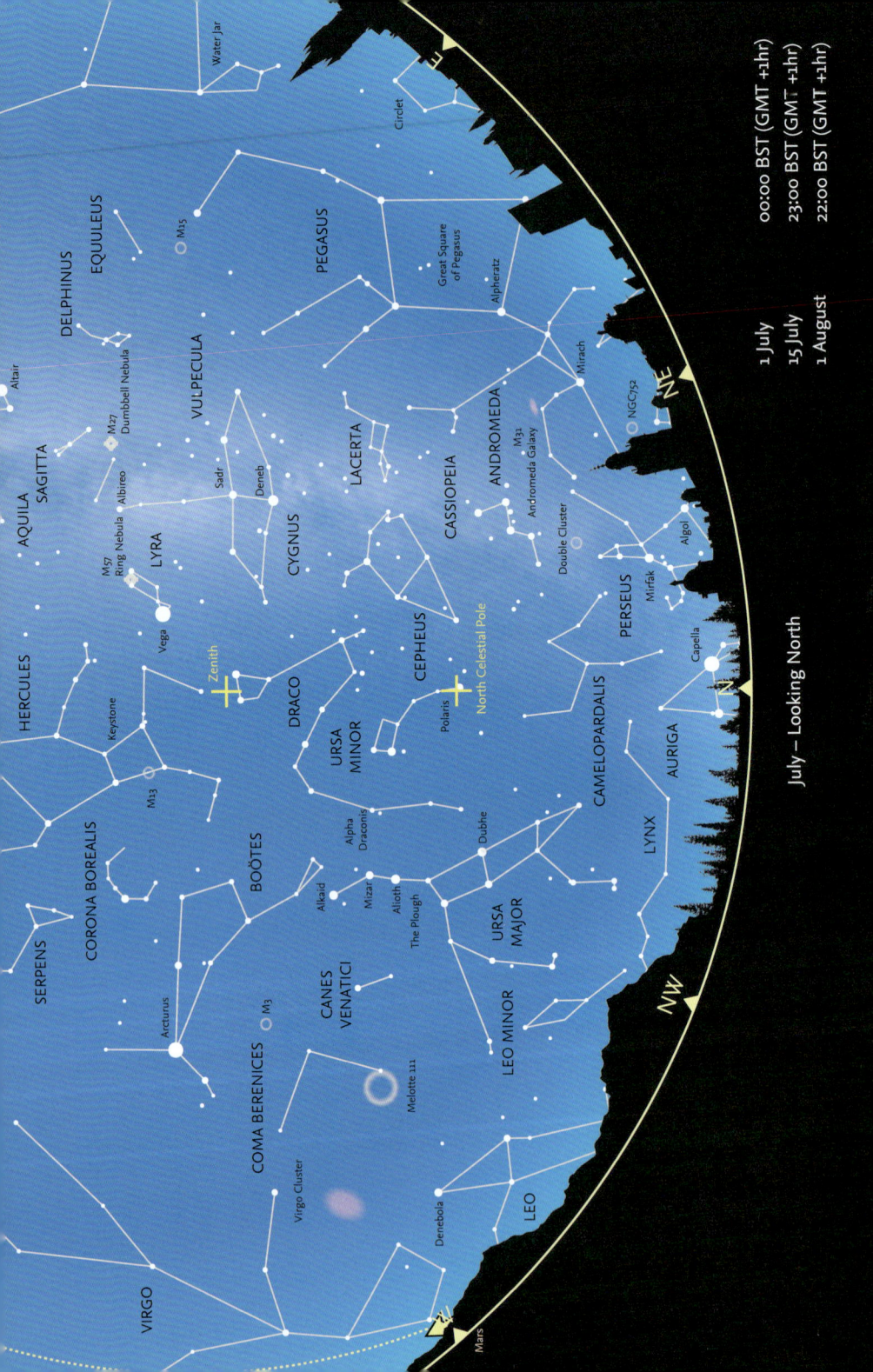

July – Looking North

As in June, light nights and the chance of observing noctilucent clouds persist throughout July, but later in the month (and particularly after midnight) some of the major constellations are more easily seen. **Capella**, the brightest star in **Auriga** (most of which is too low to be visible), is skimming the northern horizon. **Cassiopeia** is clearly visible in the northeast and **Perseus**, to its south, is beginning to climb clear of the horizon. The band of the Milky Way, from Perseus through Cassiopeia towards **Cygnus**, stretches up into the northeastern sky. The star **Albireo** (β Cygni) marks the Swan's beak and through a telescope it can be seen to be an optical double: the brighter star is an orange giant 100 times more luminous than the Sun; the second star is blue-white and 230 times more luminous than the Sun. The stars are not thought to be orbiting each other (in a binary system).

If the sky is dark and clear, you may be able to make out the small, faint constellation of **Lacerta**, lying across the Milky Way between Cassiopeia and Cygnus. In the east, the stars of **Pegasus** are now well clear of the horizon, with the main line of stars forming **Andromeda** roughly parallel to the horizon in the northeast. **Alpheratz** (α Andromedae) is actually the star at the northeastern corner of the Great Square of Pegasus. **Cepheus** and **Ursa Major** are on opposite sides of **Polaris** and **Ursa Minor**, in the east and west, respectively. The head of **Draco** is very close to the zenith so the whole of this winding constellation is readily seen.

Meteors

July brings three meteor showers; there are two minor radiants active in the constellations of **Capricornus** and **Aquarius**. Because of their location, however, observing conditions are not particularly favourable for northern-hemisphere observers. The first shower, the α-**Capricornids**, active from 3 July–15 August (peaking 30 July), has a maximum rate of about five per hour; however, it does often produce very bright fireballs. The parent body is Comet 169P/NEAT. The most prominent shower is probably that of the **Southern δ-Aquariids**, which are active from around 12 July–23 August, with a peak on 30 July, although even then the rate is unlikely to reach 25 meteors per hour. In this case, the parent body may be Comet 96P/Machholz. This year, both shower maxima occur when the Moon is a bright waning gibbous, meaning moonlight will interfere with observations. A chart showing the δ-**Aquariids** radiant is shown on page 30. The Perseids begin on 17 July and peak on 13 August.

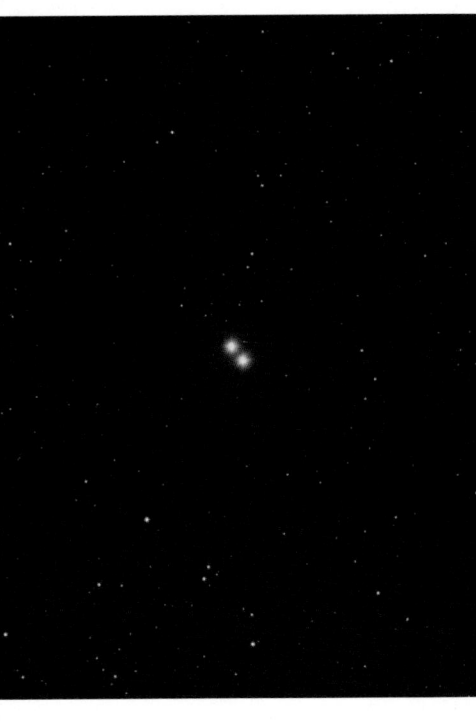

The double star ν Draconis.

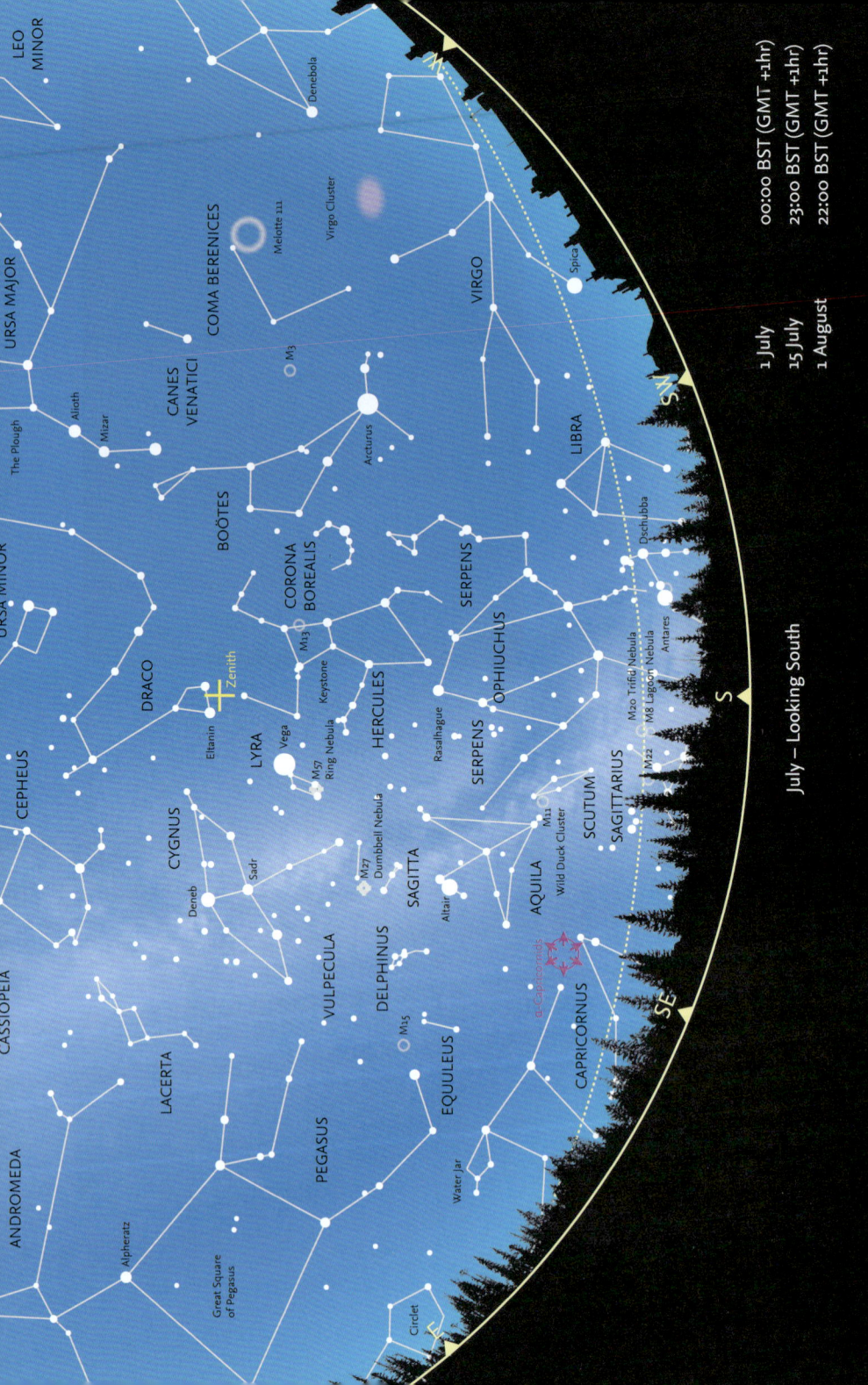

July – Looking South

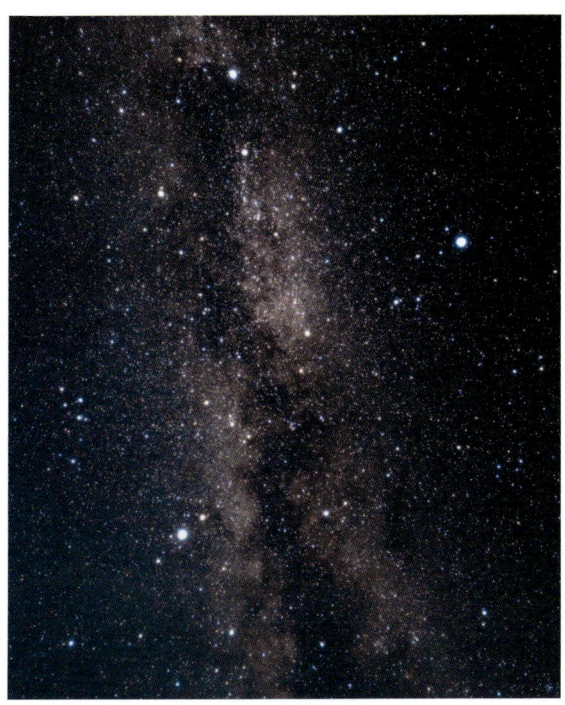

Although part of the constellation remains hidden, this is perhaps the best time of year to see **Scorpius**, with deep red **Antares** (α Scorpii), glowing just above the southern horizon. At around midnight (UT, 01:00 BST), part of **Sagittarius**, with the distinctive asterism of the Teapot, and the dense star clouds of the centre of the Milky Way, are just visible in the south. The Great Dark Rift – actually dust clouds that hide the more distant stars – runs down the Milky Way from **Cygnus** towards Sagittarius. Towards its northern end is the small constellation of **Sagitta** and the planetary nebula **M27** (the Dumbbell Nebula) in the otherwise insignificant constellation of **Vulpecula**. The sprawling constellation of **Ophiuchus** lies close to the meridian for a large part of the month, separating the two halves of the constellation of **Serpens**.

In the east, the bright Summer Triangle, consisting of **Vega** in **Lyra**, **Deneb** in **Cygnus** and **Altair** in **Aquila**, begins to dominate the southern sky, as it will throughout August and into September. The small constellation of Lyra, containing Vega (the fifth brightest star in the night sky) and a distinctive quadrilateral of stars to its east and south, lies not far south of the zenith.

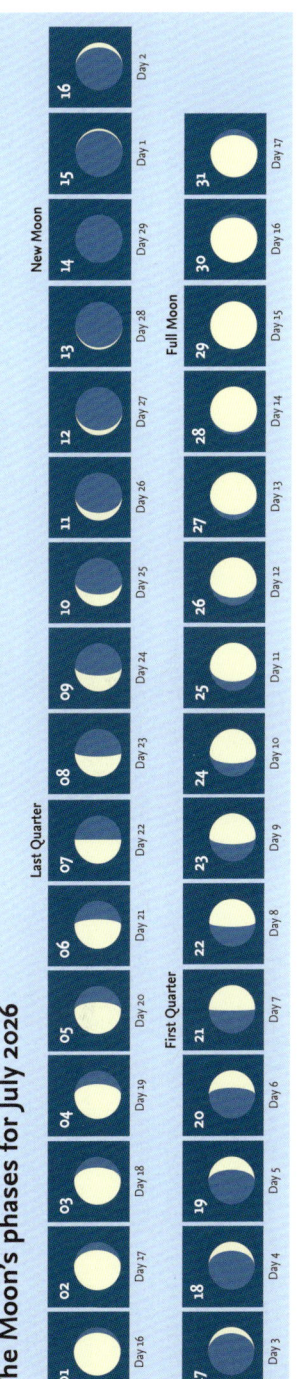

The Summer Triangle marked by the stars Deneb, Vega and Altair and the Milky Way.

The Moon's phases for July 2026

JULY 73

July – Moon and Planets

The Moon

The Moon is Last Quarter on 7 July. On 10 July the waning crescent Moon is 1.1°N of the **Pleiades**. A New Moon takes place 14 July. Three days later the waxing crescent moon passes 0.5°S of **Regulus** and 2.0°S of **Venus** (mag. -4.2). On 21 July the First Quarter Moon lies 2.4°S of **Spica**. On 24 July the waxing gibbous Moon approaches **Antares**, sitting 0.6°S of the red supergiant star. A Full Moon dominates the night sky on 29 July in **Capricornus**.

The Planets

Mercury is placed in **Gemini** and eventually appears just after sunset until the middle of the month, after which it leads the Sun at sunrise (mag. 2.2 to 5.6 then brightening again to 0.5). **Venus** stays in **Leo** for the month, appearing after sunset (mag. -4.1 to -4.3). On 9 July it passes 0.9°N of **Regulus**, visible until a few hours before midnight. **Mars** sits in **Taurus**, appearing a few hours before sunrise (mag. 1.4 to 1.3). On 4 July **Mars** is 0.1°S of **Uranus** (mag. 5.8). **Jupiter** lies in **Cancer**, visible from sunset until around 20:00. The gas giant approaches the Sun and leads the Sun in the dawn sky after 29 July (mag. -1.8). **Saturn** continues in **Pisces**, climbing above the eastern horizon after midnight (mag. 0.8 to 0.6). Saturn enters retrograde motion on 26 July at mag. 0.7. Uranus in Taurus is lost in the glare of the Sun at sunrise (mag. 5.8). **Neptune** in Pisces appears after midnight and enters retrograde motion on 7 July (mag. 7.8 to 7.7).

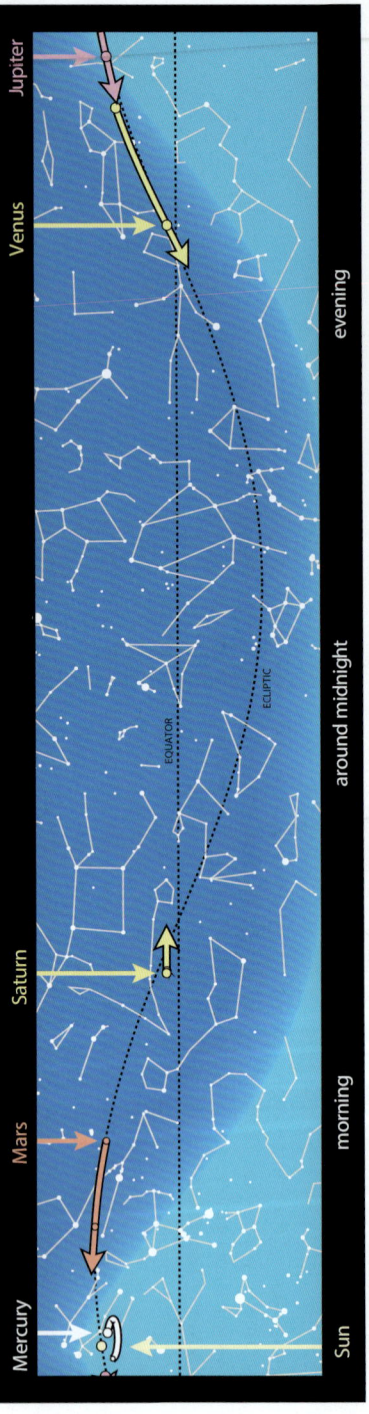

The path of the Sun and the planets along the ecliptic in July.

Calendar for July

06	17:30	Earth at aphelion (152,087,778 km = 1.01664 AU)
07	19:29	Last Quarter Moon
09	14:36	Venus (mag. -4.1) 0.9°N of Regulus
10	22:54	Pleiades 1.1°S of the Moon
13	07:50	Moon at perigee = 359,111 km
14	09:43	New Moon
17	00:07	Regulus 0.5°N of the Moon
17	16:31	Venus (mag. -4.2) 2.0°N of the Moon
21	03:21	Spica 2.4°N of the Moon
21	11:06	First Quarter Moon
24	21:00	Antares 0.6°N of the Moon
25	16:45	Moon at apogee = 405,549 km
28–29		Delta Aquariid meteor shower maximum
29	14:36	Full Moon
30		Alpha Capricornid meteor shower maximum

8 July • Last Quarter Moon in Pisces with Saturn, rising in the east after midnight.

11 July • The waning crescent Moon close to the Pleiades, Uranus and Mars. Aldebaran, Capella and Mirfak (Algenib) are nearby.

17 July • Waxing crescent Moon south of Venus and Regulus.

24 July • Waxing gibbous Moon south of Antares. Sabik and Saik of Ophiuchus sit at higher altitudes.

JULY

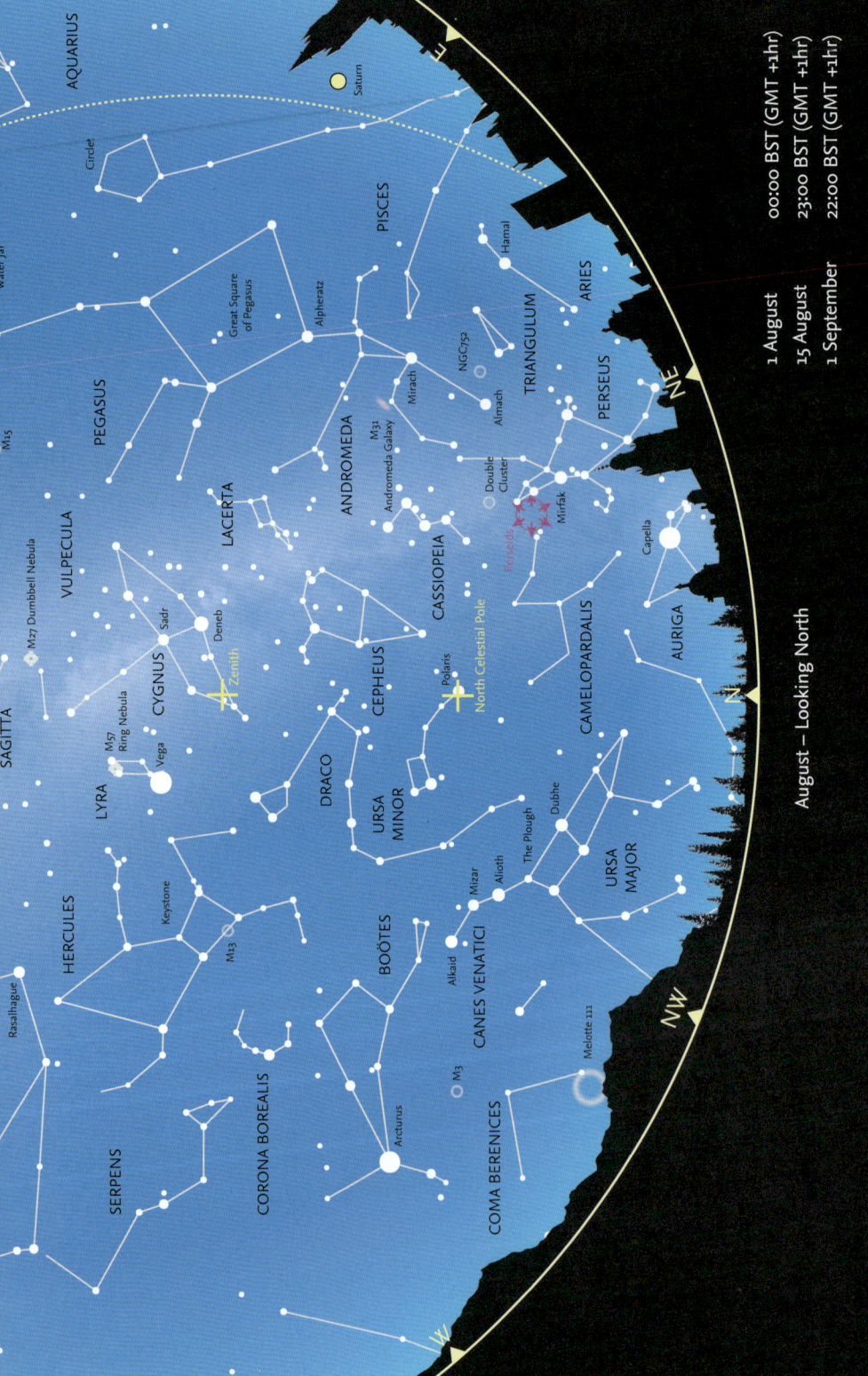

August – Looking North

Ursa Major is now the 'right way up' in the northwest, although some of the fainter stars in the south of the constellation are difficult to see. Beyond it, *Boötes* stands almost vertically in the west, but pale orange *Arcturus* is sinking towards the horizon. Higher in the sky, both *Corona Borealis* and *Hercules* are clearly visible, along with *Draco*.

In the northeast, *Capella* is clearly seen, but most of *Auriga* still remains below the horizon. Higher in the sky, *Perseus* is gradually coming into full view and, later in the night and later in the month, the beautiful *Pleiades* cluster rises above the northeastern horizon. Between Perseus and *Polaris* lies the faint constellation of *Camelopardalis*.

Higher still, both *Cassiopeia* and *Cepheus* are well placed for observation, despite the fact that Cassiopeia is completely immersed in the band of the Milky Way, as is the 'base' of Cepheus. *Pegasus* and *Andromeda* are now well above the eastern horizon and, below them, the constellation of *Pisces* is climbing into view. Two of the stars in the Summer Triangle, *Deneb* and *Vega*, are close to the zenith high overhead.

Meteors

August is the month when one of the best meteor showers of the year occurs: the *Perseids*. This is a long shower, generally beginning about 17 July and continuing until around 24 August, the peak of the shower occurs over the night of 12–13 August, when the rate may reach as high as 100 meteors per hour (and on rare occasions, even higher). In 2026, there is a New Moon at the maximum, so there will be no competition from moonlight. The Perseids are debris from Comet 109P/Swift-Tuttle (the Great Comet of 1862). Perseid meteors are fast and many of the brighter ones leave persistent trains. Some bright fireballs also occur during the shower, arising from larger chunks of cometary debris.

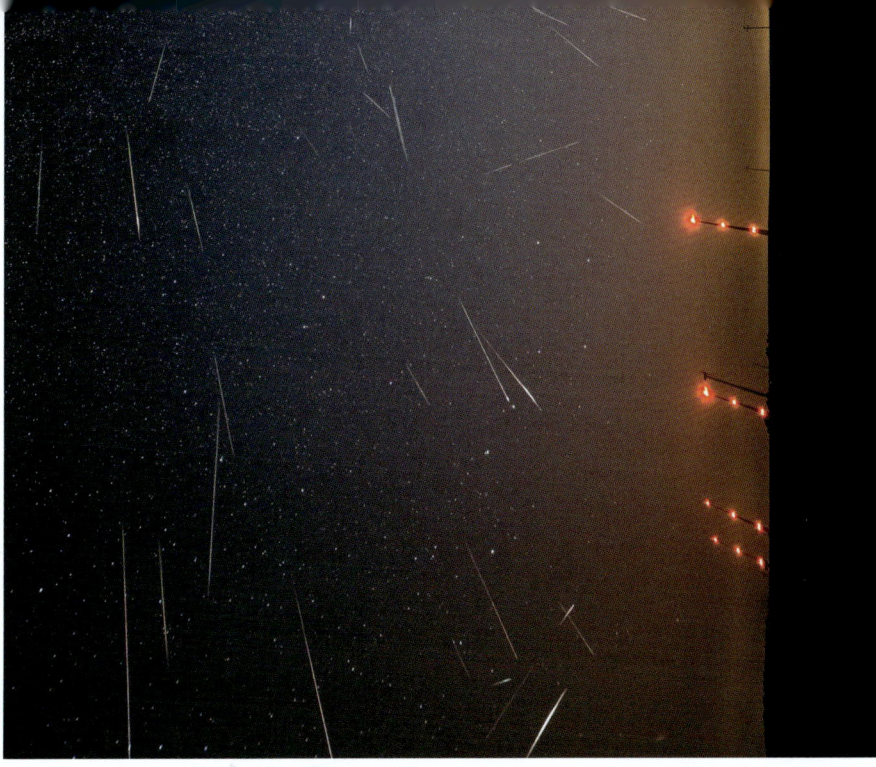

The Perseid meteor shower – a compilation of images.

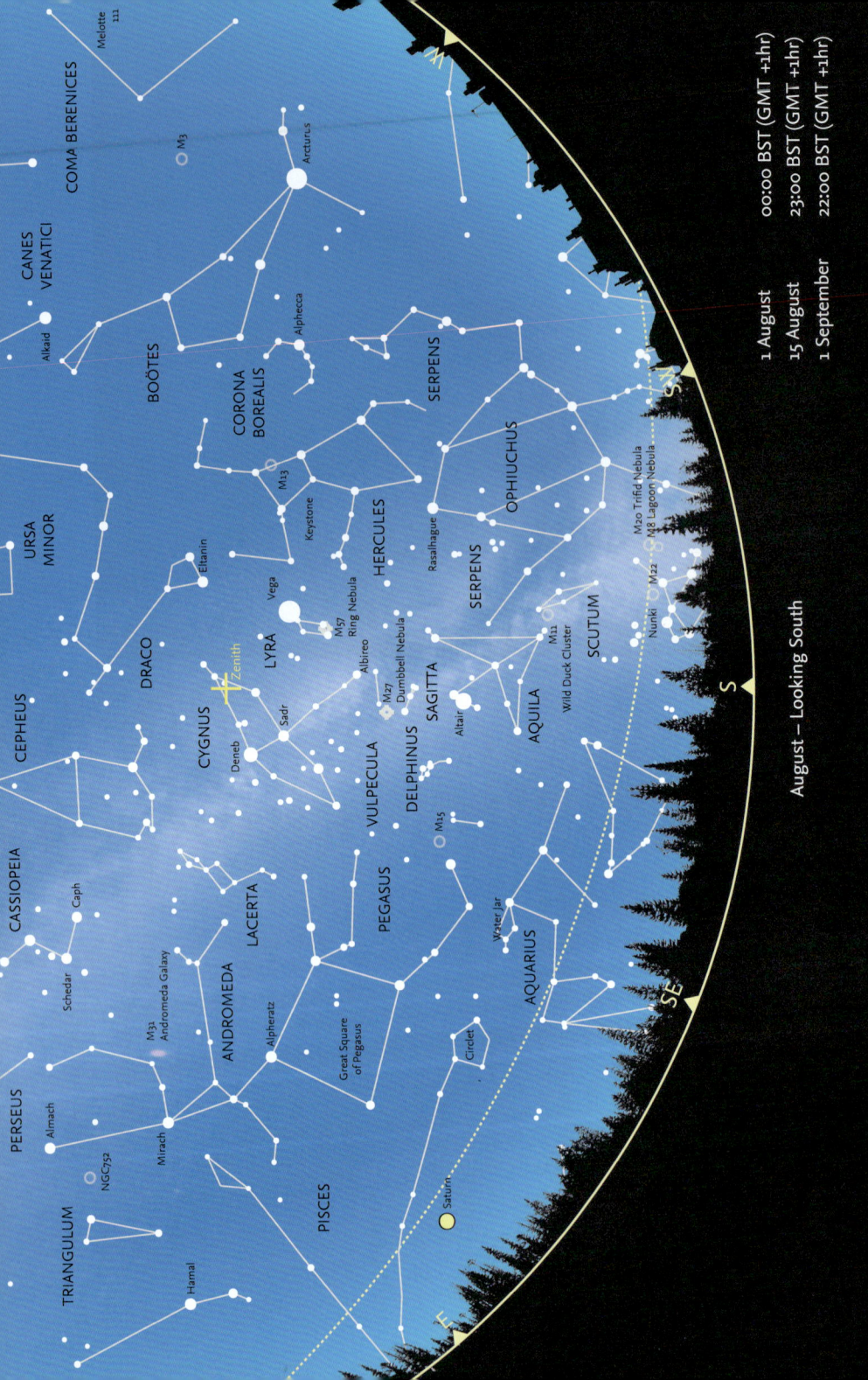

August – Looking South

The whole of the summer Milky Way stretches across the sky in the south, from **Cygnus**, high in the sky near the zenith, past **Aquila**, with bright **Altair** (α Aquilae), to part of the constellation of **Sagittarius** close to the horizon, where the pattern of stars known as the Teapot is visible. This area contains many nebulae and both open and globular clusters. The star-forming Lagoon Nebula, **M8** is close to γ Sagittarii and it is visible to the naked eye on a clear dark night. **M20**, the Trifid Nebula is close to M8 and can be seen with a small telescope.

Between **Albireo** (β Cygni) and Altair lie the two small constellations of **Vulpecula** and **Sagitta**, with the latter easier to distinguish (because of its shape) from the clouds of the Milky Way. Between Sagitta and **Pegasus** to the east lie the highly distinctive five stars that form the tiny constellation of **Delphinus** (the Dolphin), again, one of the few constellations that actually bears some resemblance to the creature after which it is named. Below Aquila, mainly in the star clouds of the Milky Way, lies **Scutum**, most famous for the bright open cluster, **M11** or the Wild Duck Cluster, readily visible in binoculars. To the southeast of Aquila lie the two zodiacal constellations of **Capricornus** and **Aquarius**.

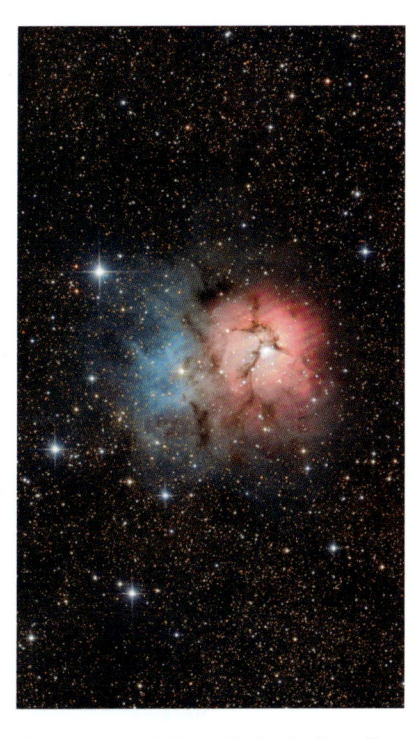

The Trifid Nebula (M20), a hydrogen emission cloud in a star-forming region of the Milky Way.

The Moon's phases for August 2026

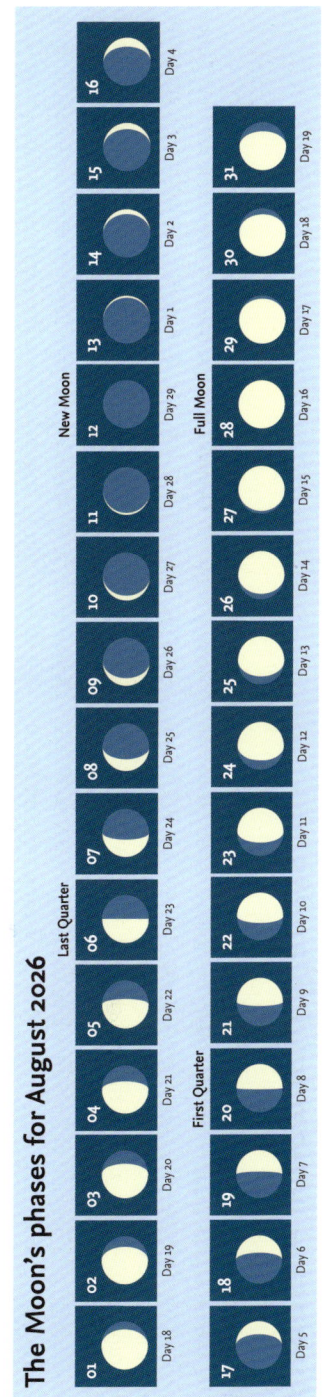

August – Moon and Planets

The Moon

The Moon is Last Quarter on 6 August; the following day it passes 1.2°N of the **Pleiades**. On 9 August **Mars** (mag. 1.3) is 4.4°S of the Moon; a day later the waning crescent Moon is 3.6°S of **Pollux**. On 11 August **Mercury** (mag. -0.9) is 2.1°S of the thin crescent Moon. A partial solar eclipse will be visible from London on 12 August at 17:46, at maximum obscuration 91.4 per cent of the Sun will be covered by the New Moon. On 16 August **Venus** (mag. -4.4) lies 2.1°N of the waxing crescent Moon in the early evening. The following day the Moon approaches **Spica**, the star is 2.4°N. On 20 August the Moon is First Quarter; a day later it passes 0.6°S of **Antares**. On 28 August a partial lunar eclipse will be visible low on the western horizon from the UK, maximum coverage occurs at 04:13.

The Planets

Mercury begins in **Gemini**, visible at dawn. A week later it enters **Cancer** and ends the month in **Leo**. After 27 August it trails the Sun, appearing shortly after sunset. Its magnitude brightens from 0.3 to -1.9 (at solar conjunction) and then dims to -1.6. It reaches western elongation on 2 August (mag. 0.2). On 15 August Mercury (mag. -1.2) is 0.5°N of **Jupiter** (mag. -1.8). **Venus** moves from Leo into **Virgo** on 1 August (mag. -4.3 to -4.6). On 15 August Venus is at eastern elongation (mag. -4.4). **Mars** starts in **Taurus** and swings into Gemini in the middle of the month, it brightens from 1.4 to 1.2 and dims slightly to 1.3 at the end of the month. Mars appears 1.5°N of (1) **Ceres** on 9 August in Taurus. Jupiter can be seen just before sunrise in Cancer (mag. -1.8). **Saturn** is in **Pisces**, visible after 23:00 (mag. 0.6 to 0.5). **Uranus** lies in Taurus, above the horizon from around midnight (mag. 5.8 to 5.7). **Neptune** is in Pisces and visible after 22:00 (mag. 7.7).

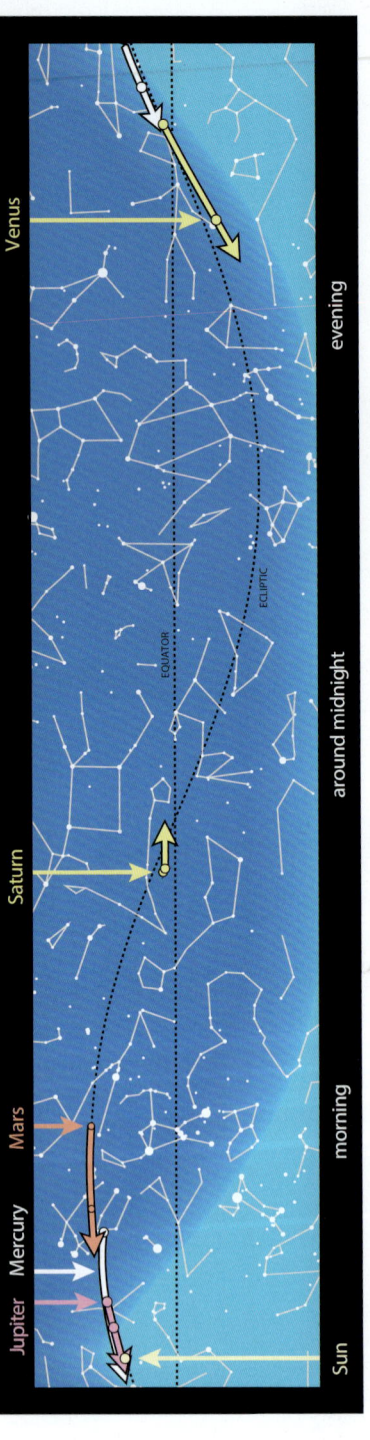

The path of the Sun and the planets along the ecliptic in August.

Calendar for August

02	08:00	Mercury at greatest elongation (19.5°W, mag. 0.2)
06	02:21	Last Quarter Moon
07	06:23	Pleiades 1.2°S of the Moon
09	05:31	Mars (mag. 1.3) 4.4°S of the Moon
10	11:18	Moon at perigee = 363,288 km
10	22:38	Pollux 3.6°N of the Moon
11	12:48	Mercury (mag. -0.9) 2.1°S of the Moon
12	17:37	New Moon
12	17:46	Total solar eclipse (Iceland, Spain, Greenland & the Arctic). Partial eclipse visible from North America, western Africa & Europe
12–13		Perseid meteor shower maximum
15	06:00	Venus at greatest elongation (45.9°E, mag. -4.4)
16	08:47	Venus (mag. -4.4) 2.1°N of the Moon
17	11:49	Spica 2.4°N of the Moon
20	02:46	First Quarter Moon
21	04:18	Antares 0.6°N of the Moon
22	08:20	Moon at apogee = 404,644 km
28	04:13	Partial lunar eclipse. Visible from Europe (including London), Africa & the eastern Pacific
28	04:18	Full Moon

6–9 August · The Last Quarter Moon on 6 August, moving past the Pleiades and Mars as a waning crescent.

12 August · Partial solar eclipse as seen from London, UK. Maximum cover (91%) at 19:13 BST.

15–17 August · The waxing crescent Moon and Venus at greatest eastern elongation on 15 August. Moon next to Spica on 17 August.

28 August · Lunar eclipse visible from London at 05:13 BST.

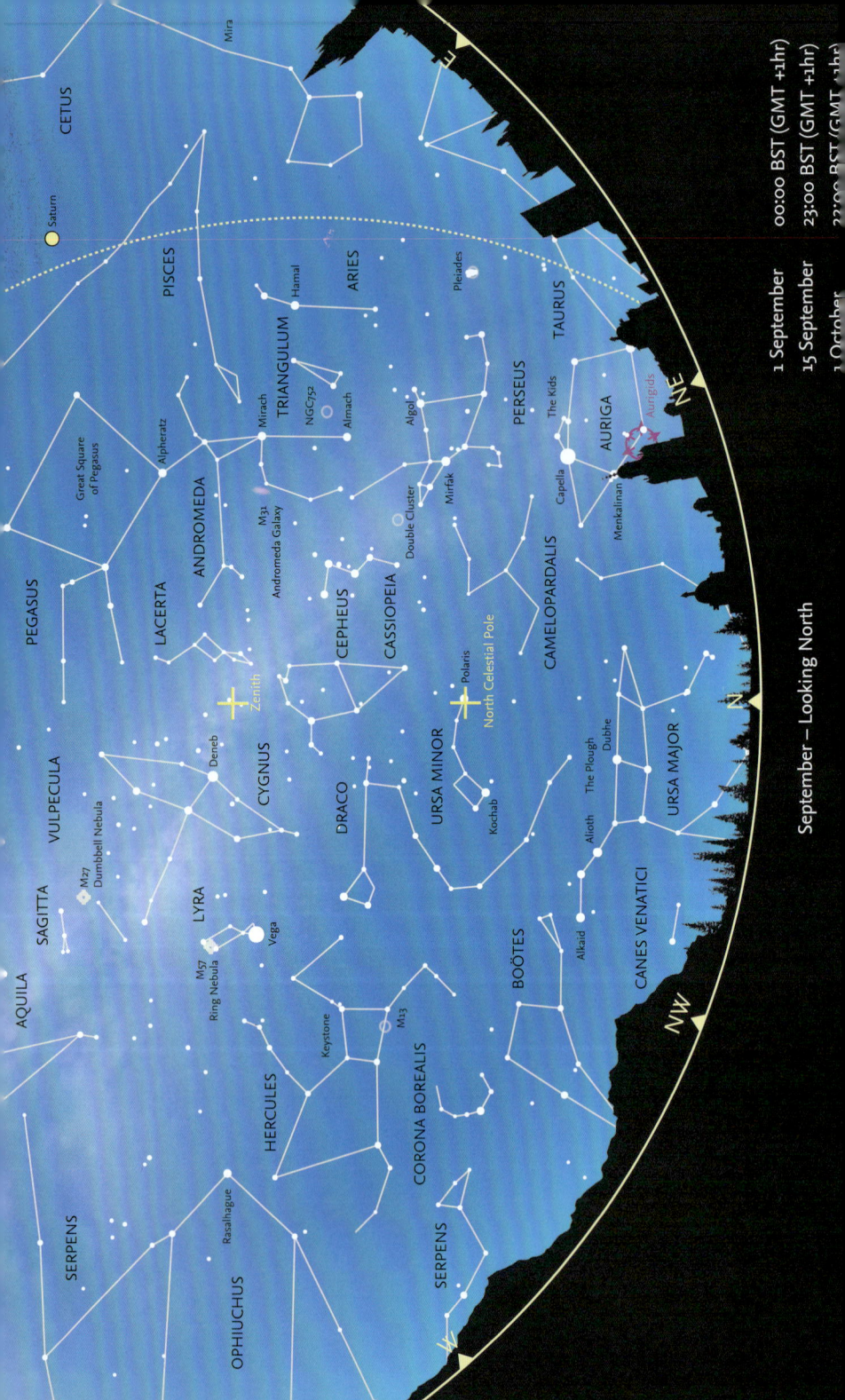

September – Looking North

The (northern) autumnal equinox occurs on 23 September, when the Sun moves south of the equator in Virgo.

Ursa Major is now low in the north and to the northwest **Arcturus** and much of **Boötes** sink below the horizon later in the night and later in the month. In the northeast, **Auriga** is beginning to climb higher in the sky. Later in the month, **Taurus**, with orange **Aldebaran** (α Tauri), and even **Gemini**, with **Castor** and **Pollux**, become visible in the east and northeast. Due east, **Andromeda** is now clearly visible, with the small constellations of **Triangulum** and **Aries** (the latter a zodiacal constellation) directly below it. Practically the whole of the northern Milky Way is visible, arching across the sky, both in the north and in the south. The clouds of stars are easier to see in **Cassiopeia** and **Cygnus**. The **Double Cluster** in Perseus is well placed for observation. **Cepheus** is 'upside down' near the zenith, and the head of **Draco** and **Hercules** (with the globular cluster **M13**) beyond it are clearly visible.

Meteors

There are two minor showers in September. The α-**Aurigids** starts 28 August and ends 5 September and tends to have two peaks of activity. The principal peak occurs on 1 September. In 2026, the Moon will be a bright waning gibbous, so conditions are unfavourable during the shower. At maximum, however, the hourly rate hardly reaches 10 meteors per hour, although the meteors are bright and relatively easy to photograph. The **Southern Taurid** shower begins this month (on 10 September) and, although rates are low, often produces very bright fireballs. This is a very long shower, lasting until about 20 November.

M13, the Hercules Globular Cluster, contains several hundred thousand stars. First discovered by Edmond Halley in 1714.

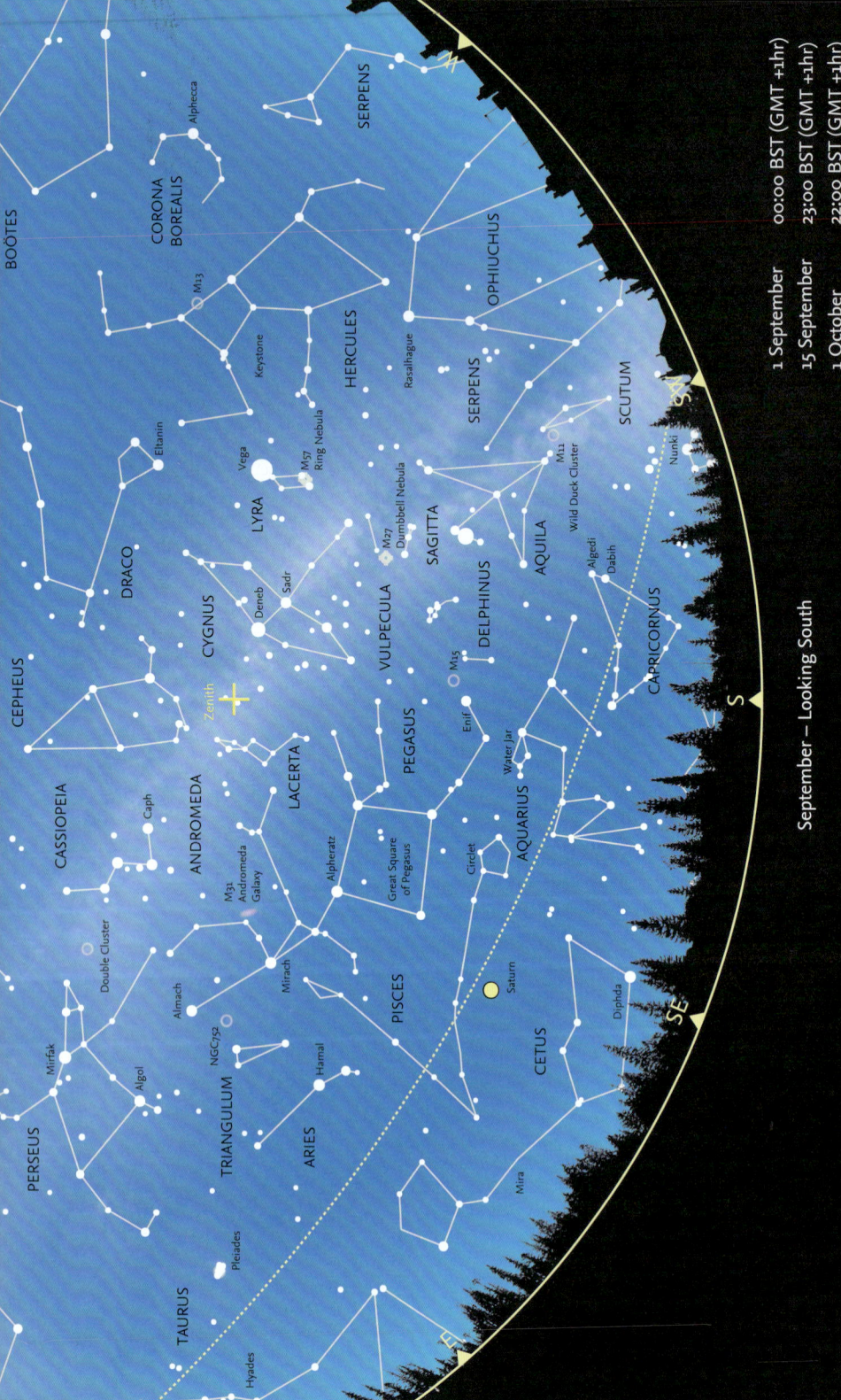

September – Looking South

The Summer Triangle is now high in the southwest, with the Great Square of **Pegasus** high in the southeast. Below Pegasus are the two zodiacal constellations of **Capricornus** and **Aquarius**. **Algedi** (α Capricorni) is actually an optical double, with the two stars (α^1 Cap and α^2 Cap) readily seen with the naked eye. **Dabih** (β Capricorni), just to the south, is also a double star and the components are relatively easy to separate with binoculars. In Aquarius, there is a small asterism consisting of four stars, resembling a tiny letter 'Y', known as the Water Jar. Below Aquarius is a sparsely populated area of the sky with just one bright star in the constellation of **Piscis Austrinus**. In classical illustrations, water is shown flowing from the Water Jar towards bright **Fomalhaut** (α Piscis Austrini).

Another zodiacal constellation, **Pisces**, is now clearly visible to the east of Aquarius. Although faint, there is a distinctive asterism of stars, known as the Circlet, south of the Great Square and another line of faint stars to the east of Pegasus. Still farther down towards the horizon is the constellation of **Cetus**, with the famous variable red giant star **Mira** (o Ceti) at its centre. When Mira is at maximum brightness (around mag. 3.5) it is clearly visible to the naked eye, but it disappears as it fades towards minimum (about mag. 9.5 or less).

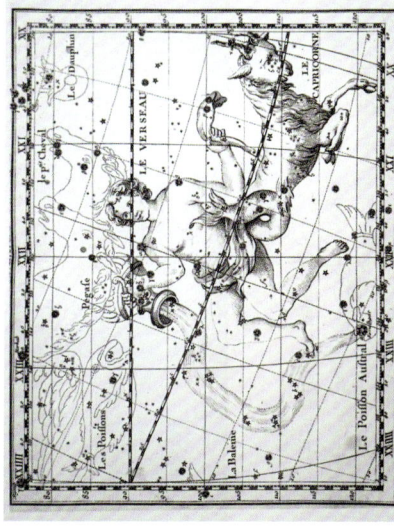

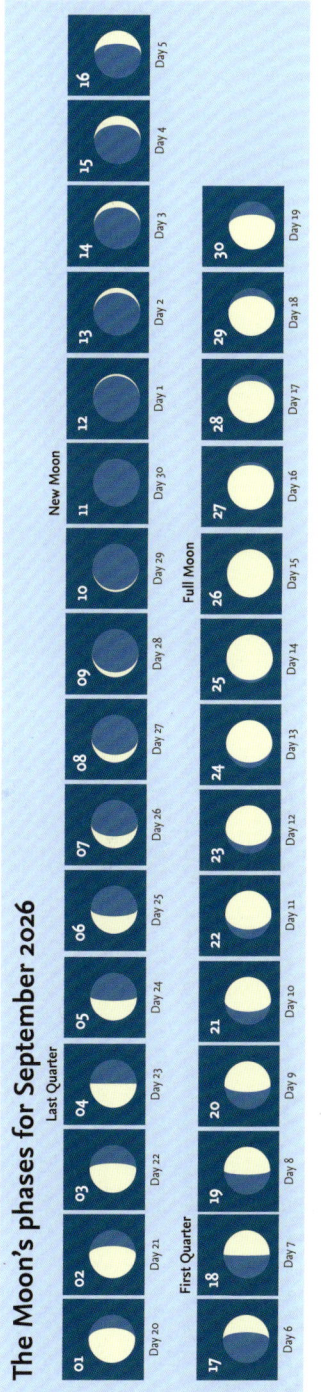

Aquarius including the Water Jar asterism and Fomalhaut (α Piscis Austrini).

The Moon's phases for September 2026

September – Moon and Planets

The Moon

On 3 September the waning gibbous Moon sits 1.2°N of the *Pleiades*. The following day it reaches Last Quarter. On 6 September *Mars* (mag. 1.2) is 3.0°S of the waning crescent Moon; a day later it moves 3.6°S of *Pollux*. On 8 September the Moon moves 0.8°N of *Jupiter* (mag. -1.8); the following day the thin crescent Moon is 0.5°S of *Regulus*. The New Moon takes place on 11 September; two days later the faint waxing crescent Moon is 2.4°S of *Spica*. *Venus* (mag. -4.8) is 0.5°S of the Moon on 14 September; an occultation will occur in the morning, lost in the daylight. On 17 September *Antares* lies 0.6°N of the waxing crescent Moon, followed by a First Quarter Moon a day later. A Full Moon occurs on 26 September and the Moon rejoins the Pleiades on 30 September, lying 1.1°N of the star cluster.

The Planets

Mercury starts in *Leo* and enters *Virgo* by the end of the first week of September, visible after sunset (mag. -1.5 to -0.1). On 26 September it moves 0.8°N of *Spica*, at mag. -0.2. *Venus* sits in Virgo for the month, approaching the Sun and visible in the early evening (mag. -4.6 to -4.8). On 1 September it is 1.2°S of Spica. *Mars* starts the month in *Gemini* and slowly moves into *Cancer*. The red planet is visible from around 02:00 until sunrise (mag. 1.2 to 1.1). *Jupiter* can be observed a few hours before sunrise in Cancer, crossing into Leo near the end of the month (mag. -1.8 to -1.9). *Saturn* lies on the border between *Pisces* and *Cetus*, visible after 21:00 (mag. 0.5 to 0.3). *Uranus* stays in *Taurus*, visible in the late evening (mag. 5.7). It enters retrograde motion on 10 September. *Neptune* is in Pisces (mag. 7.7), reaching opposition on 26 September, a distance of 28.9 AU from Earth.

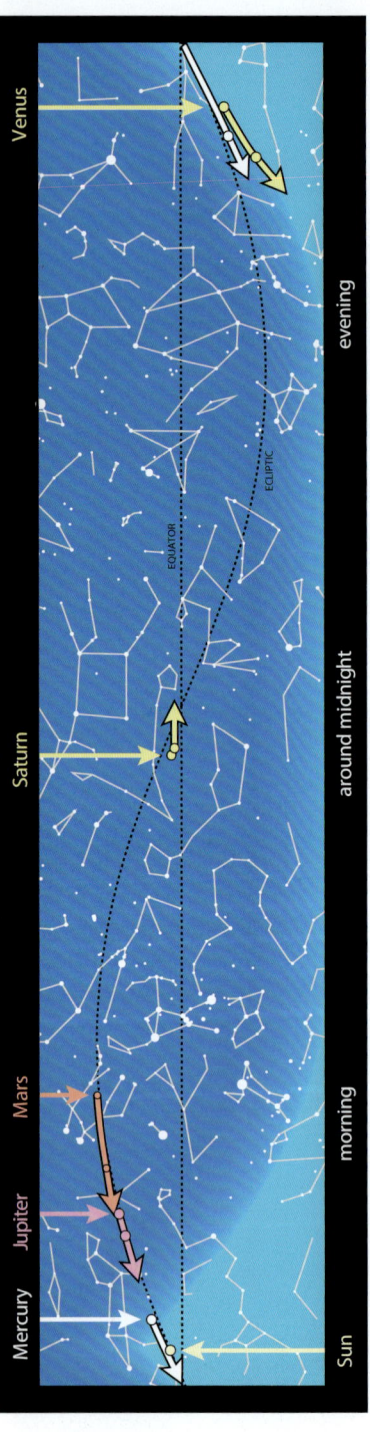

The path of the Sun and the planets along the ecliptic in September.

Calendar for September

01	13:24	Venus (mag. -4.6) 1.2°S of Spica
03	12:03	Pleiades 1.2°S of the Moon
04	07:51	Last Quarter Moon
06	18:24	Mars (mag. 1.2) 3.0°S of the Moon
06	20:26	Moon at perigee = 368,255 km
07	06:32	Pollux 3.6°N of the Moon
08	18:13	Jupiter (mag. -1.8) 0.8°S of the Moon
09	19:36	Regulus 0.5°N of the Moon
11	03:27	New Moon
13	20:53	Spica 2.4°N of the Moon
14	11:10	Venus (mag. -4.8) 0.5°S of the Moon Occultation visible from Asia, Africa, Europe (including London) & western Russia
17	12:18	Antares 0.6°N of the Moon
18	20:44	First Quarter Moon
19	03:00	Moon at apogee = 404,217 km
23	00:06	Autumnal Equinox
26	01:49	Mercury (mag. -0.2) 0.8°N of Spica
26	16:49	Full Moon

3 September • *Waning gibbous Moon next to the Pleiades, Uranus nearby, along with Aldebaran, Capella, Elnath and Betelgeuse.*

6 September • *Waning crescent Moon approaching Mars in Gemini, Pollux and Castor close by.*

9 September • *Waning crescent Moon and Jupiter, Asellus Australis (δ Cnc) in Cancer and Algieba (γ Leo) in Leo nearby.*

14 September • *Waxing crescent Moon and Venus just after sunset. An occultation takes place at 11:10 UTC.*

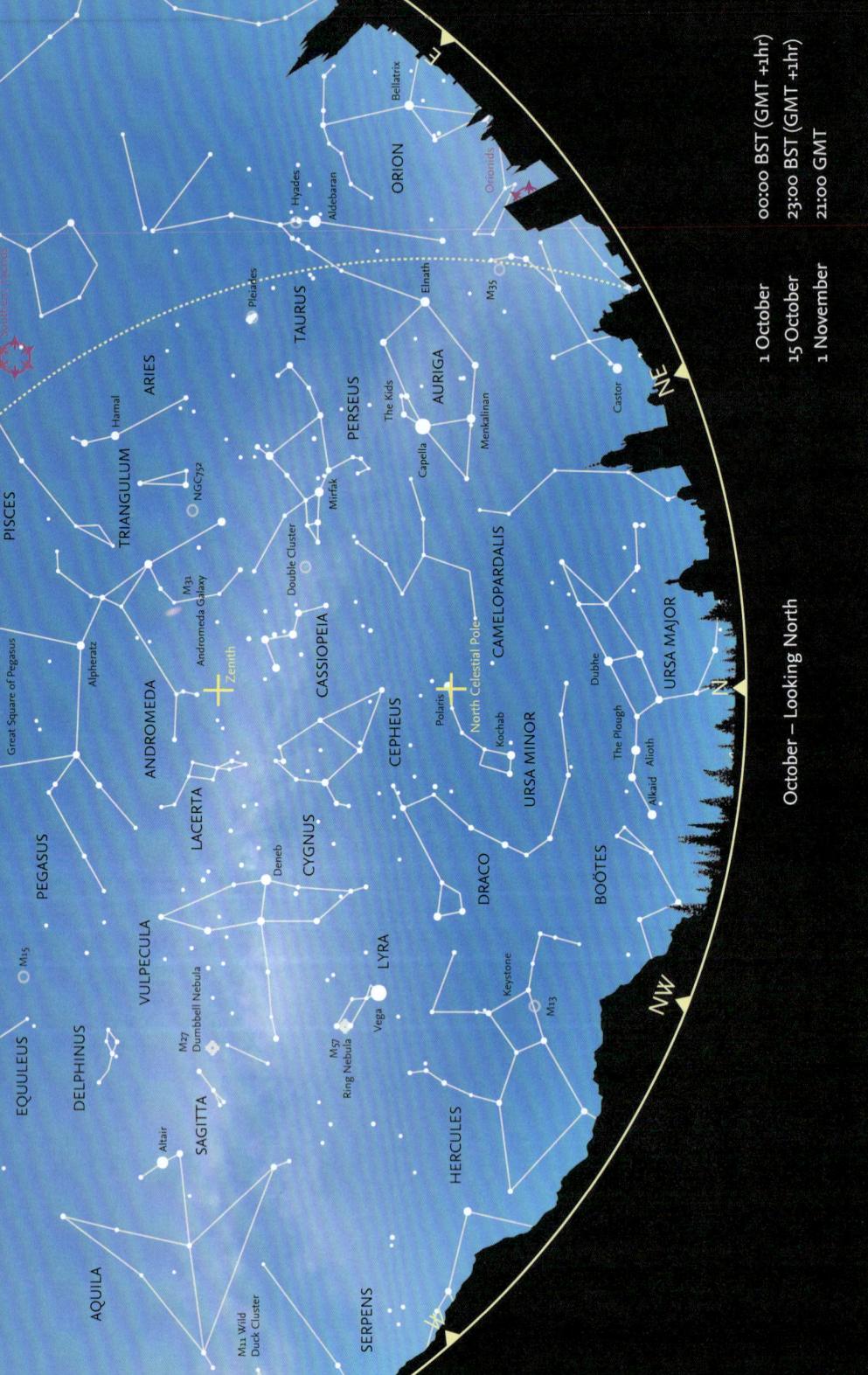

October – Looking North

Ursa Major is grazing the horizon in the north, while high overhead are the constellations of **Cepheus**, **Cassiopeia** and **Perseus**, with the Milky Way between Cepheus and Cassiopeia near the zenith. **Auriga** is now clearly visible in the east, as is **Taurus** with the **Pleiades**, **Hyades** and orange **Aldebaran**. Also in the east, **Orion** and **Gemini** are starting to rise clear of the horizon.

The constellations of **Boötes** and **Corona Borealis** are now essentially lost to view in the northwest, and **Hercules** is also descending towards the western horizon. The three stars of the Summer Triangle are still clearly visible, although **Aquila** and **Altair** are beginning to approach the horizon in the west. Towards the end of the month (Sunday 25 October) Summer Time ends in Europe, with Britain reverting to Greenwich Mean Time and Europe to Central European Time.

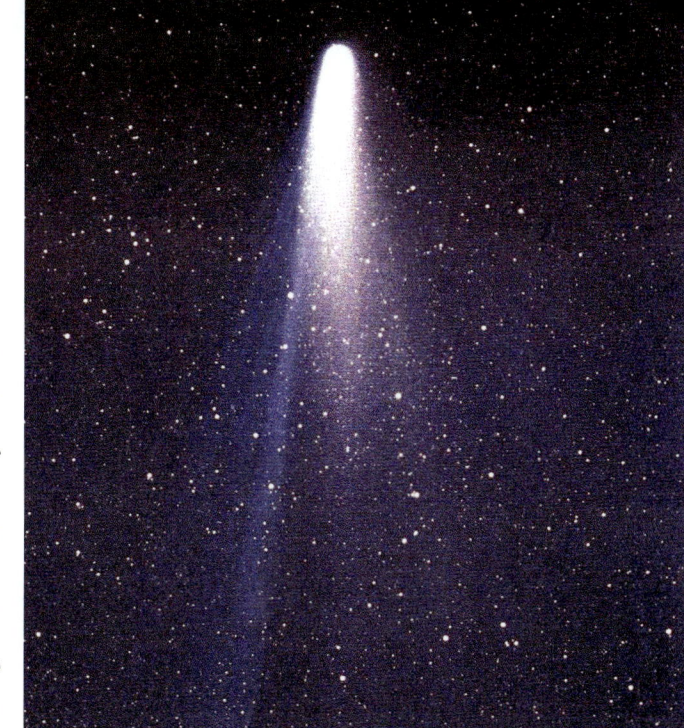

bright. A minor shower of short duration, the **Draconids**, begins on 6 October and peaks on 9 October, when skies are dark as the thin waning crescent Moon rises just before dawn.

Comet 1P/Halley imaged in March 1986. Debris from the comet provide the source of the Orionids meteor shower.

Meteors

The **Orionids** meteor shower starts on 2 October and builds up to a broad maximum lasting around a week centred on 21 October. Like the May η-**Aquariid** shower, the Orionids are associated with Comet 1P/Halley. During this second pass through the stream of particles from the comet, slightly fewer meteors are seen than in May, but conditions are more favourable for northern observers. In both showers the meteors are very fast and many leave persistent trains, with hourly rates around 15. In 2026, there is a waxing gibbous Moon lighting up the sky and interfering with observations.

The faint shower of the **Southern Taurids** (often with bright fireballs) peaks on 10 October when the Moon is New and observing conditions are favourable. Towards the end of the month (around 20 October), another shower (the **Northern Taurids**) begins to show activity, peaking in November. The parent comet for both Taurid showers is Comet 2P/Encke. The meteors in both Taurid streams are relatively slow and

OCTOBER 89

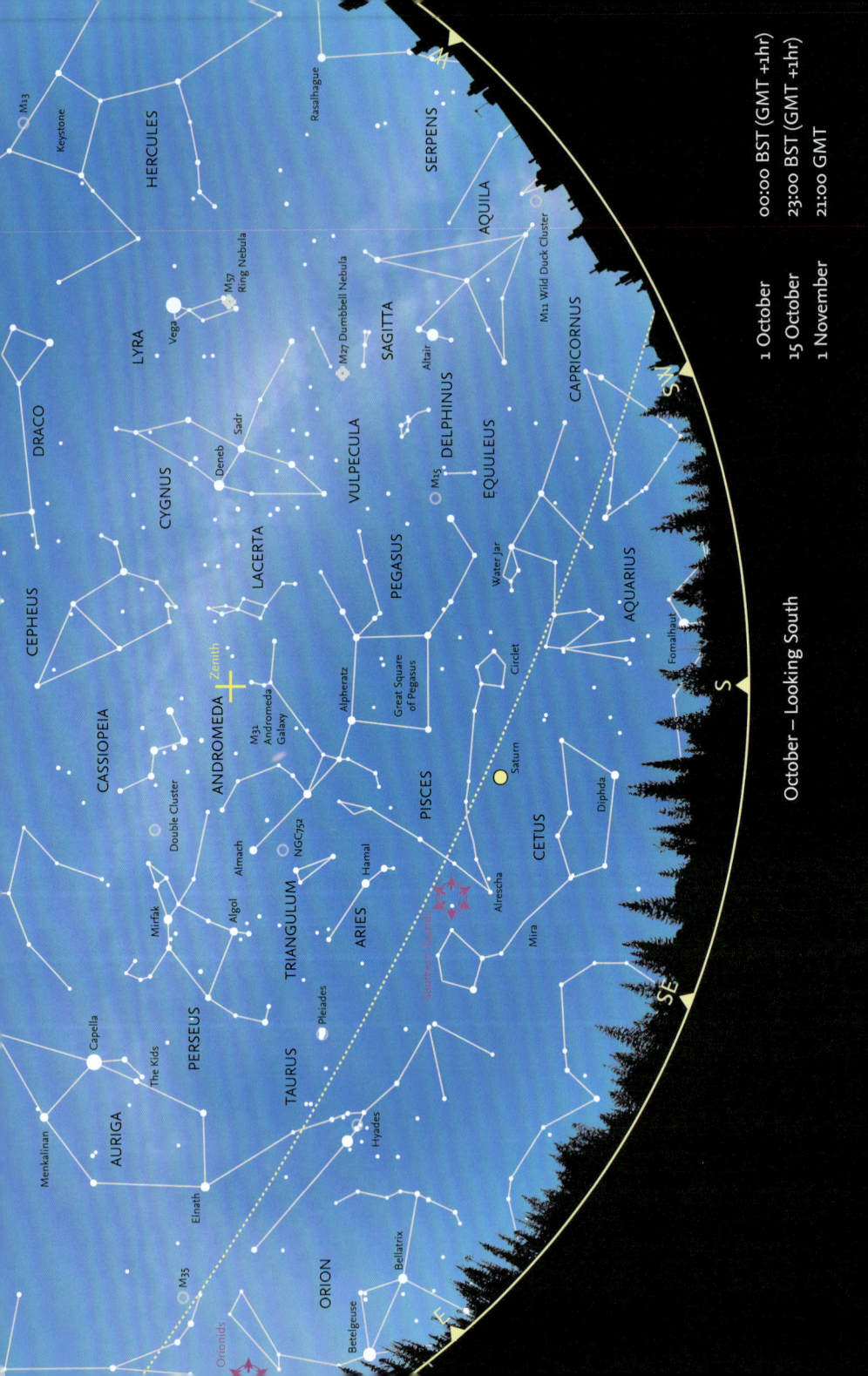

October – Looking South

The Great Square of *Pegasus* dominates the southern sky, framed by the two chains of stars that form the constellation of *Pisces*, together with *Alrescha* (α Piscium) at the point where the two lines of stars join. Also clearly visible is the constellation of *Cetus*, below Pegasus and Pisces. Although *Capricornus* is beginning to disappear, *Aquarius* to its east is well placed in the south, with solitary *Fomalhaut* and the constellation of *Piscis Austrinus* (the Southern Fish) beneath it, close to the horizon.

The main band of the Milky Way and the Great Dark Rift runs down from *Cygnus*, through *Vulpecula*, *Sagitta* and *Aquila* towards the western horizon. *Delphinus* and the tiny, unremarkable constellation of *Equuleus* lie between the band of the Milky Way and Pegasus. *Andromeda* is clearly visible high in the sky to the southeast, with the small constellation of *Triangulum* and the zodiacal constellation of *Aries* below it. *Perseus* is high in the east, and by now the *Pleiades* and *Taurus* are well clear of the horizon. Later in the night, and later in the month, *Orion* rises in the east, a sign that the autumn season has arrived, and winter is approaching.

Fomalhaut, the brightest star of Piscis Austrinus (the Southern Fish).

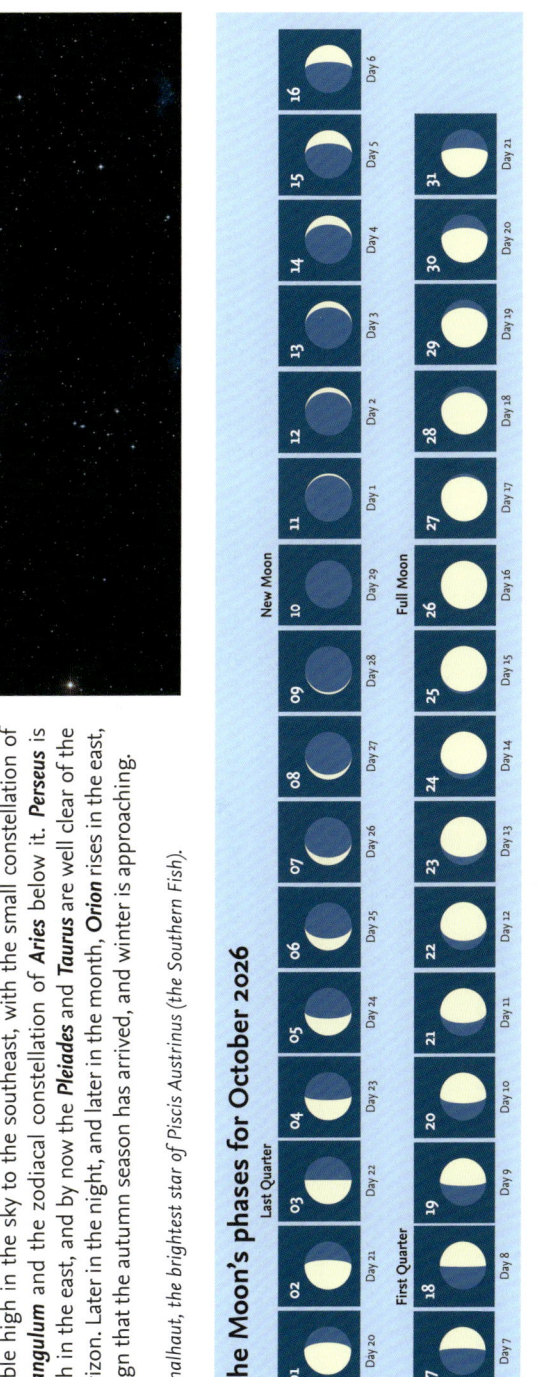

The Moon's phases for October 2026

OCTOBER

October – Moon and Planets

The Moon

On 3 October the Last Quarter Moon dominates the late night sky. A day later the waning crescent Moon sits 3.8°S of *Pollux*. On 5 October *Mars* (mag. 1.1) sits 1.2°S of the Moon; a day later the Moon moves 0.2°N of *Jupiter* (mag. -1.9). On 7 October *Regulus* is 0.6°N of the faint crescent Moon; a New Moon takes place on 10 October. On 12 October the thin waxing crescent Moon is 3.1°N of *Venus* (mag. -4.5) and later that day it is 2.1°S of *Mercury* (mag. 0.0). Two days later the Moon joins *Antares*, sitting 0.4°S of the red star. On 18 October the Moon is First Quarter; a Full Moon occurs on 26 October. Two days later it lies 1.0°N of the *Pleiades* in *Taurus*. The waning gibbous Moon rejoins *Pollux* on 31 October, sitting 4.0°S of one of the heads of *Gemini*.

The Planets

Mercury begins the month in *Virgo* and swiftly moves into *Libra*. It is visible after sunset and approaches the Sun after eastern elongation on 12 October (mag. -0.1 to 2.6). *Venus* is in *Virgo* visible in the early evening. It approaches the Sun and is seen at dawn after 24 October (mag. -4.8 to -4.1). *Mars* is in *Cancer*, it progresses into *Leo* for Halloween, approaching *Jupiter* along the ecliptic and visible after midnight (mag. 1.1 to 0.8). On 11 October Mars (mag. 1.1) sits next to the Beehive Cluster, M44 (mag. 3.1), 1.0°N. Jupiter is in Leo, climbing above the horizon from 01:00 (mag. -1.9 to -2.0). *Saturn* lies on the border between *Pisces* and *Cetus*, visible after sunset (mag. 0.3 to 0.5). On 4 October it is at opposition (mag. 0.3), positioned 8.4 AU from Earth. *Uranus* is in *Taurus* and visible in the late evening (mag. 5.7 to 5.6). *Neptune* is in Pisces and visible just after sunset (mag. 7.7).

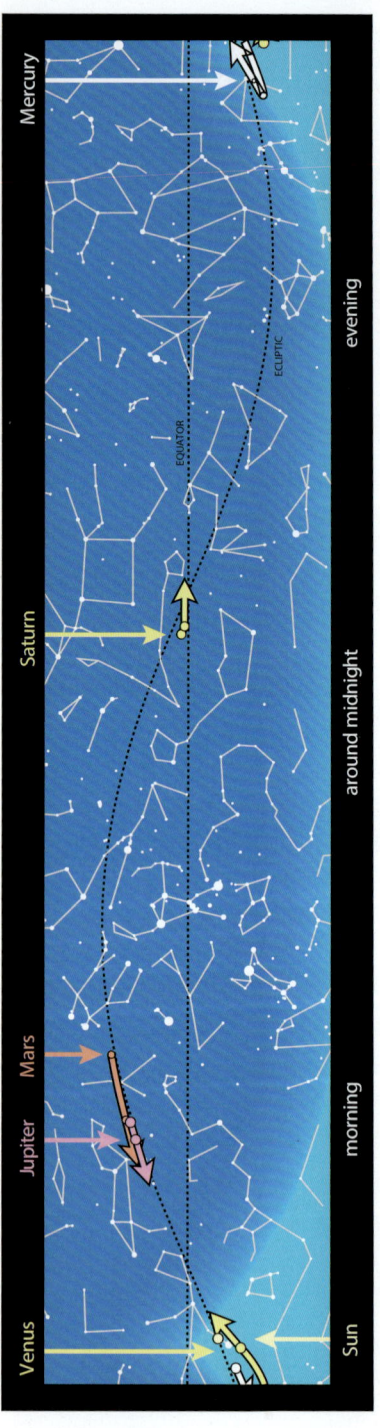

The path of the Sun and the planets along the ecliptic in October.

Calendar for October

01	20:41	Moon at perigee = 369,338 km
03	13:25	Last Quarter Moon
04	12:21	Saturn (mag. 0.3) at opposition
04	12:27	Pollux 3.8°N of the Moon
05	05:30	Mars (mag. 1.1) 1.2°S of the Moon
06	10:18	Jupiter (mag. -1.9) 0.2°S of the Moon
07	02:57	Regulus 0.6°N of the Moon
10	15:50	New Moon
10		Southern Taurid meteor shower maximum
12	02:30	Venus (mag. -4.5) 3.1°S of the Moon
12	10:00	Mercury at greatest elongation (25.2°E, mag. 0.0)
12	20:08	Mercury (mag. 0.0) 2.1°N of the Moon
14	20:25	Antares 0.4°N of the Moon
16	22:56	Moon at apogee = 404,639 km
18	16:13	First Quarter Moon
21–22		Orionid meteor shower maximum
26	04:12	Full Moon
28	01:11	Pleiades 1.0°S of the Moon
28	18:01	Moon at perigee = 364,411 km

3 October • Last Quarter Moon in Gemini, Mars in adjacent Cancer. Pollux, Castor, Betelgeuse and Procyon are visible.

4 October • Saturn in Pisces. Diphda, Alpheratz, Algenib and Markab nearby.

5 October • Waning crescent Moon next to Mars, Jupiter sits nearby, Procyon in the vicinity.

31 October • Waning gibbous Moon close to Pollux and Castor, Betelgeuse and Procyon are nearby.

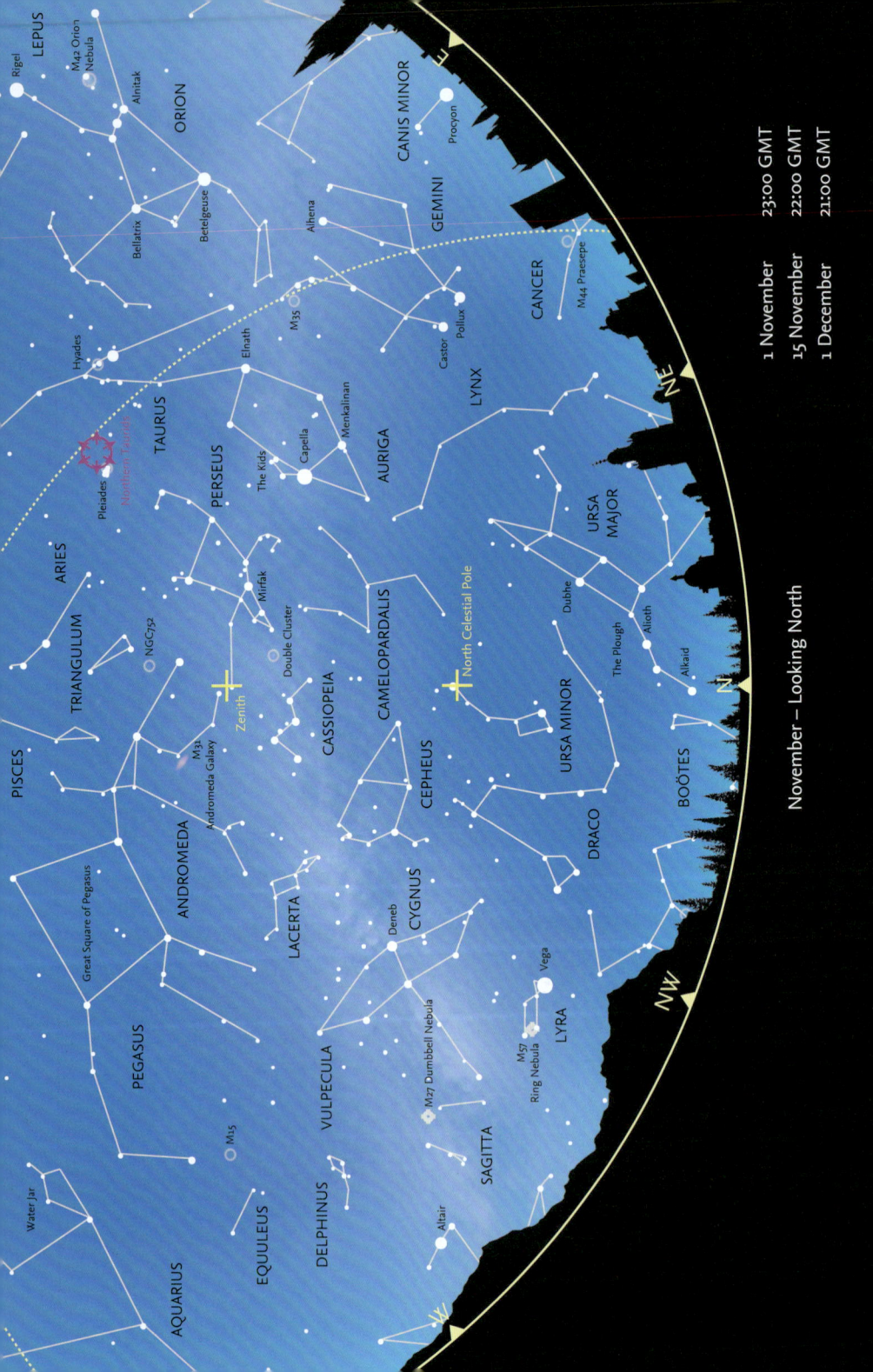

November – Looking North

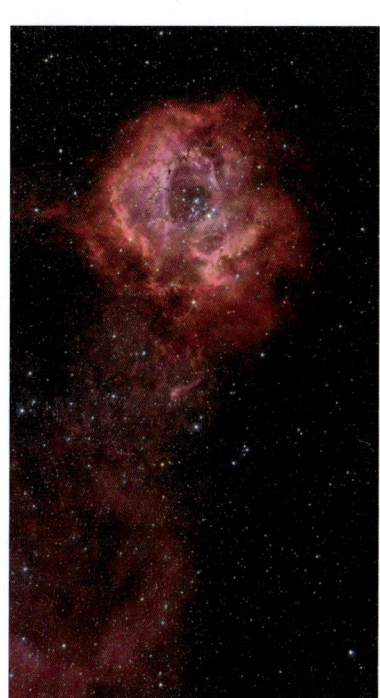

The Rosette Nebula, a hot glowing hydrogen gas cloud in Monoceros surrounding a cluster of young stars considerably hotter and more luminous than the Sun.

Most of **Aquila** has now disappeared below the horizon, but two of the stars of the Summer Triangle, **Vega** in **Lyra** and **Deneb** in **Cygnus**, are still clearly visible in the west. The head of **Draco** is now low in the northwest and only a small portion of **Hercules** remains above the horizon. The southernmost stars of **Ursa Major** are now coming into view. The Milky Way arches overhead, with the denser star clouds in the west and the less heavily populated region through **Auriga** and **Monoceros** in the east. The star β **Monocerotis** is actually a triple star system; all three stars are six to seven times the mass of the Sun. High overhead, **Cassiopeia** is near the zenith and **Cepheus** has swung round to the northwest, while Auriga is now high in the northeast. **Gemini**, with **Castor** and **Pollux**, is well clear of the eastern horizon, and **Procyon** (α Canis Minoris), in reality a binary star system, is just climbing into view almost due east.

Meteors

The **Northern Taurid** shower, which began in mid-October, reaches maximum – although only with a rate of about five meteors per hour – on 12 November. The Moon is waxing crescent and setting relatively early; the meteors are best observed later in the evening. The shower gradually trails off, ending around 10 December. Far more striking, however, are the **Leonids**, which have a relatively short period of activity (6–30 November), with maximum on the evening of 17 November and early morning of 18 November. This shower is associated with Comet 55P/Tempel-Tuttle and has shown extraordinary activity on various occasions with many thousands of meteors per hour. The rate in 2026 is likely to be about 10–15 per hour at peak activity; however, the Moon will be waxing gibbous. Conditions will be favourable after midnight. These meteors are the fastest shower meteors recorded (about 70 km per second) and often leave persistent trains.

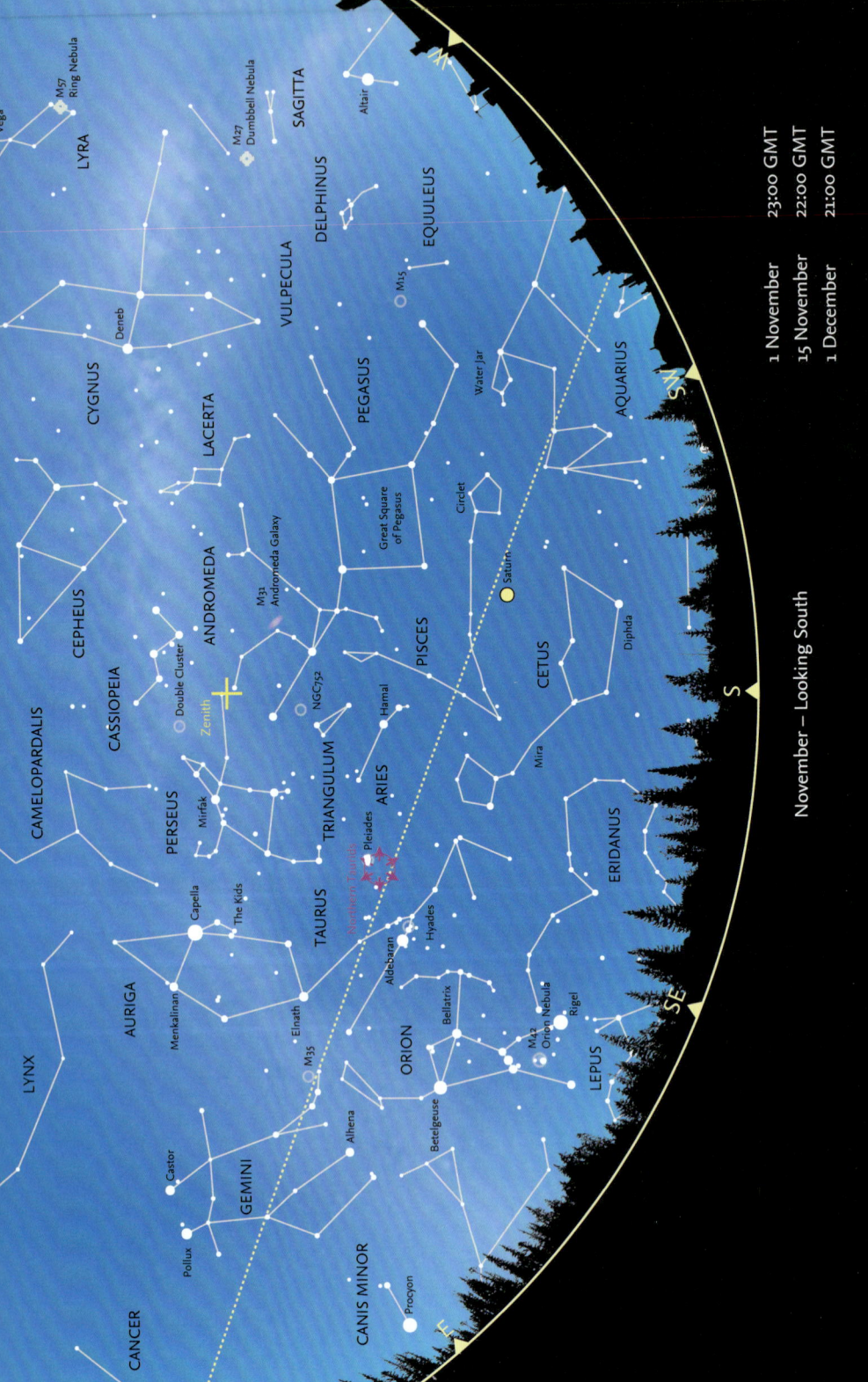

November – Looking South

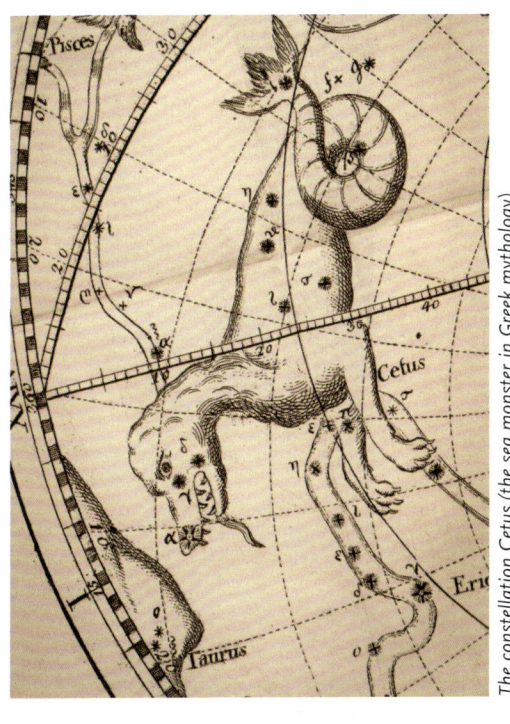

The constellation Cetus (the sea monster in Greek mythology).

Orion has now risen above the eastern horizon, and part of the long, straggling constellation of *Eridanus* (which begins near *Rigel*) is visible to the west of Orion. Higher in the sky, *Taurus*, with the *Pleiades* cluster, and orange *Aldebaran* are now easy to observe. To their west, both *Pisces* and *Cetus* are close to the meridian. The famous long-period variable star *Mira* (o Ceti), with a typical range of magnitude 3.4 to 9.5, is favourably placed for observation. In the southwest, *Capricornus* has slipped below the horizon, but *Aquarius* remains visible. Even farther west, *Altair* may be seen early in the night, but most of *Aquila* has already disappeared from view. *Delphinus*, together with *Sagitta* and *Vulpecula* in the Milky Way, will soon vanish for another year. Both *Pegasus* and *Andromeda* are easy to see, and one of the lines of stars that make up Andromeda finishes close to the zenith, which is also close to one of the outlying stars of *Perseus*, high in the east.

The Moon's phases for November 2026

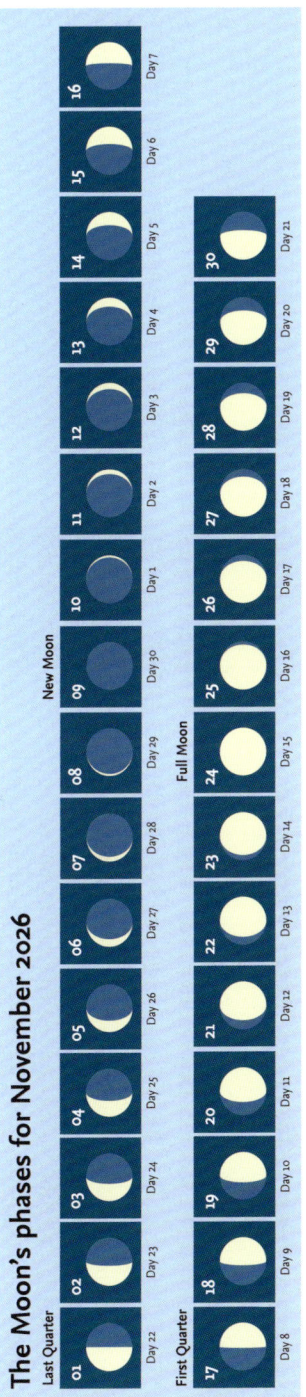

November – Moon and Planets

The Moon

On 1 November the Moon is Last Quarter; a day later the waning gibbous Moon is 1.1°S of *Mars* (mag. 0.8) and 0.5°S of *Jupiter* (mag. -2.1). On 3 November *Regulus* is 0.8°N of the waning crescent Moon. On 7 November the Moon joins *Venus* (mag. -4.5) in *Virgo*, sitting 1.1°S of the planet. It is also 2.4°S of *Spica*. On 9 November the Moon is New; two days later the slender waxing crescent Moon moves 0.3°S of *Antares*. A First Quarter Moon dominates the sky on 17 November; a week later on 24 November the Full Moon is 0.9°N of the *Pleiades*. On 28 November the waning gibbous Moon lies 4.2°S of *Pollux*; two days later it is with Jupiter (mag. -2.2), 1.2°S of the planet and 1.1°S of *Regulus* in *Leo*. On the same day Mars (mag. 0.4) is 3.3°N of the Moon.

The Planets

Mercury is placed in *Libra* close to the Sun at sunset. It leads the Sun in the west after 4 November and moves into *Virgo* and then back into Libra as the month progresses (mag. 3.3 to 6.5, brightening to -0.7). On 20 November Mercury is at western elongation (mag. -0.5). *Venus* is visible at dawn in Virgo, receding from the Sun in the sky (mag. -4.2 to -4.9). On 10 November it lies 0.1°S of *Spica* (mag. -4.7). *Mars* is in *Leo* and creeps above the horizon around an hour before midnight (mag. 0.8 to 0.4). Mars (mag. 0.7) is 1.2°N of *Jupiter* (mag. -2.1) on 16 November. On 25 November Mars (mag. 0.6) is 1.6°N of *Regulus*. *Saturn* is in Jupiter is visible in Leo from midnight (mag. -2.0 to -2.2). *Saturn* is in *Cetus* near *Pisces*, in the sky just after sunset (mag. 0.5 to 0.6). *Uranus* is in *Taurus* from early evening (mag. 5.6), reaching opposition on 25 November at a distance of 18.4 AU from Earth. *Neptune* is in Pisces (mag. 7.7).

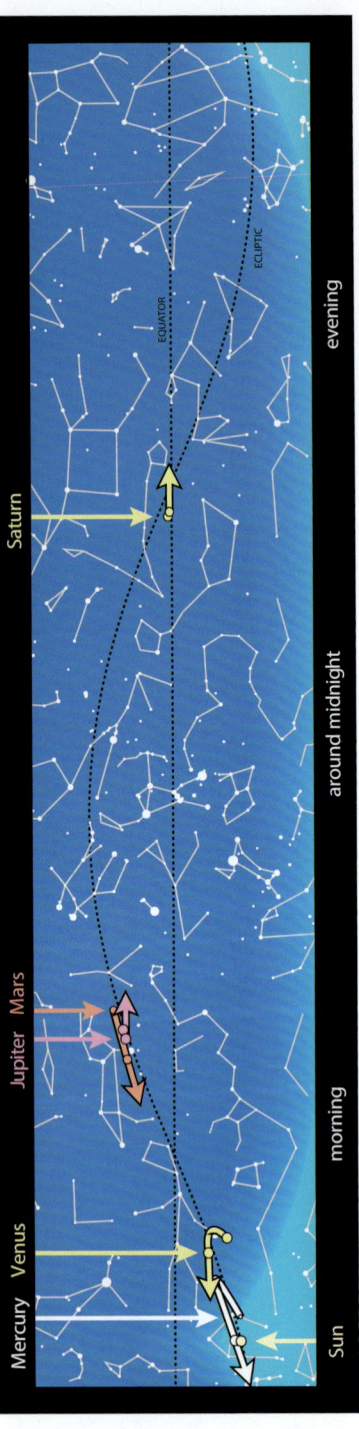

The path of the Sun and the planets along the ecliptic in November.

Calendar for November

01	20:28	Last Quarter Moon
02	14:23	Mars (mag. 0.8) 1.1°N of the Moon
02	23:11	Jupiter (mag. -2.1) 0.5°N of the Moon
03	08:40	Regulus 0.8°N of the Moon
07	11:31	Venus (mag. -4.5) 1.1°N of the Moon
07	12:40	Spica 2.4°N of the Moon
09	07:02	New Moon
10	13:49	Venus (mag. -4.7) 0.1°S of Spica
11	03:58	Antares 0.3°N of the Moon
12		Northern Taurid meteor shower maximum
13	17:50	Moon at apogee = 405,619 km
16	06:24	Mars (mag. 0.7) 1.2°N of Jupiter
17	11:48	First Quarter Moon
17–18		Leonid meteor shower maximum
20	23:40	Mercury at greatest elongation (19.6°W, mag. -0.5)
24	11:18	Pleiades 0.9°S of the Moon
24	14:53	Full Moon
25	07:47	Mars (mag. 0.6) 1.6°N of Regulus
25	20:58	Moon at perigee = 359,348 km
25	22:41	Uranus (mag. 5.6) at opposition
28	01:27	Pollux 4.2°N of the Moon
30	09:18	Jupiter (mag. -2.2) 1.2°N of the Moon
30	14:35	Regulus 1.1°N of the Moon

2 November • *The Last Quarter Moon with Mars and Jupiter, Procyon and Regulus close by.*

7 November • *Venus and the waning crescent Moon in Virgo visible shortly before sunrise. Arcturus visible in the eastern sky.*

16 November • *Mars and Jupiter in conjunction in Leo, Alphard visible in Hydra, Procyon in the western sky.*

20 November • *Mercury at western elongation, Venus close to Spica in Virgo in the dawn sky.*

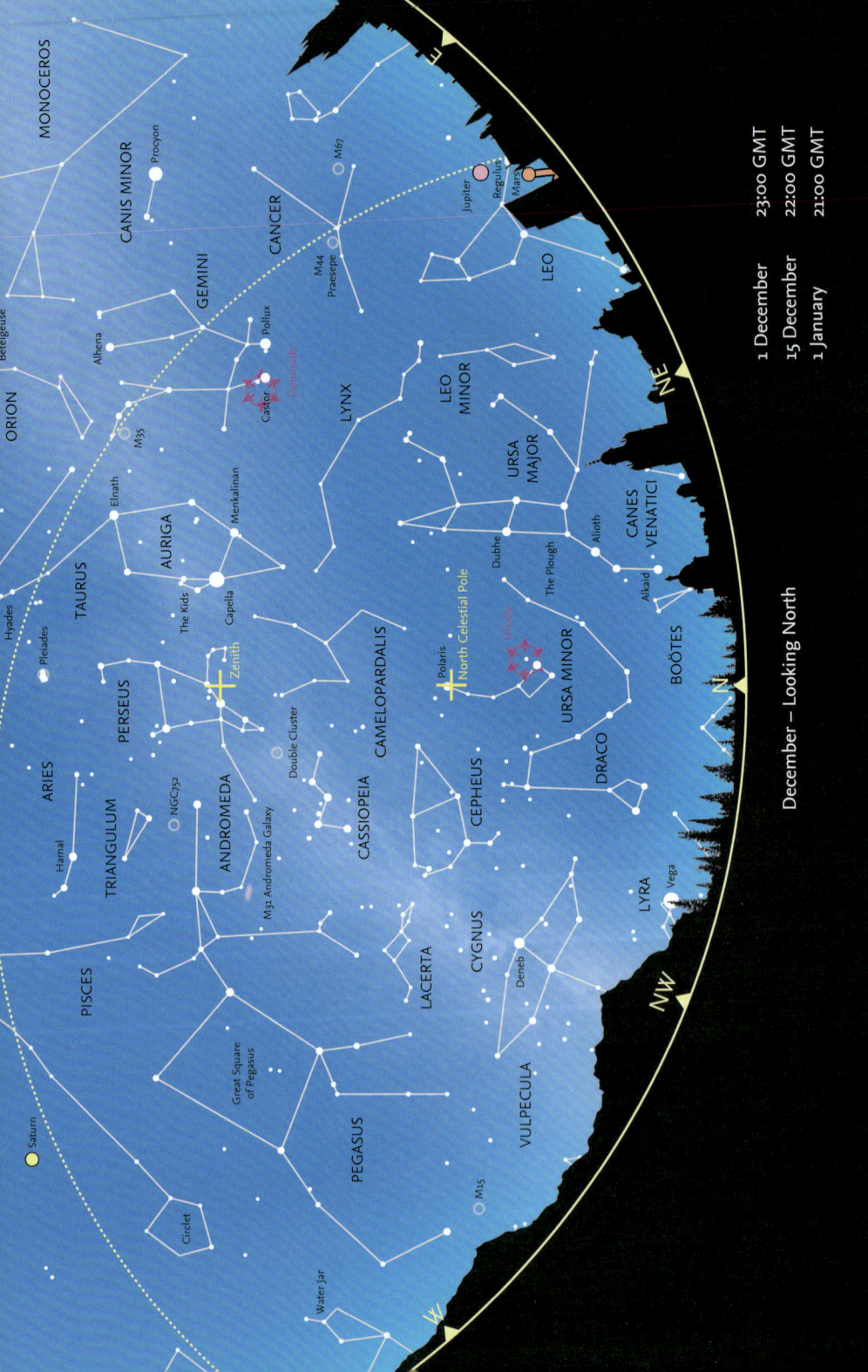

December – Looking North

The Great Bear, **Ursa Major** has now swung around and is starting to 'climb' in the east. The fainter stars in the southern part of the constellation are now fully in view. The other bear, **Ursa Minor**, 'hangs' below **Polaris** in the north. Directly above it is the faint constellation of **Camelopardalis**, with the other circumpolar constellation, **Lynx**, to its east. **Vega** (α Lyrae) is skimming the horizon in the northwest, but **Deneb** (α Cygni) and most of **Cygnus** remain visible farther west. In the east, **Regulus** (α Leonis) and the constellation of **Leo** are beginning to rise above the horizon. **Cancer** stands high in the east, with **Gemini** even higher in the sky. **Perseus** is at the zenith, with **Auriga** and **Capella** between it and Gemini. Because it is so high in the sky, now is a good time to examine the star clouds of the fainter portion of the Milky Way, between **Cassiopeia** in the west to Gemini and **Orion** in the east.

Meteors

There is one significant meteor shower in December (the last major shower of the year). This is the **Geminid** shower, which is visible over the period 4–17 December and comes to maximum on 14 December, when the Moon is a waxing crescent; favourable conditions arise later in the evening. It is one of the most active showers of the year, with a peak rate of around 100 meteors per hour. It is the one major shower that shows good activity before midnight. The source of the Geminids is debris from an asteroid called 3200 Phaethon; most meteor showers originate from cometary debris. The Geminids are assumed to consist of denser, rocky material, they are slower than most other meteors and often appear to last longer. The brightest often break up into numerous luminous fragments that follow similar paths across the sky. There is a second shower: the **Ursids**, active in the last half of the month and peaking on the 22nd, with a rate at maximum of 5–10, occasionally rising to 25 per hour. The peak in 2026 occurs when the Moon is a waxing gibbous rising late in the evening, so conditions are unfavourable. The parent body is Comet 8P/Tuttle.

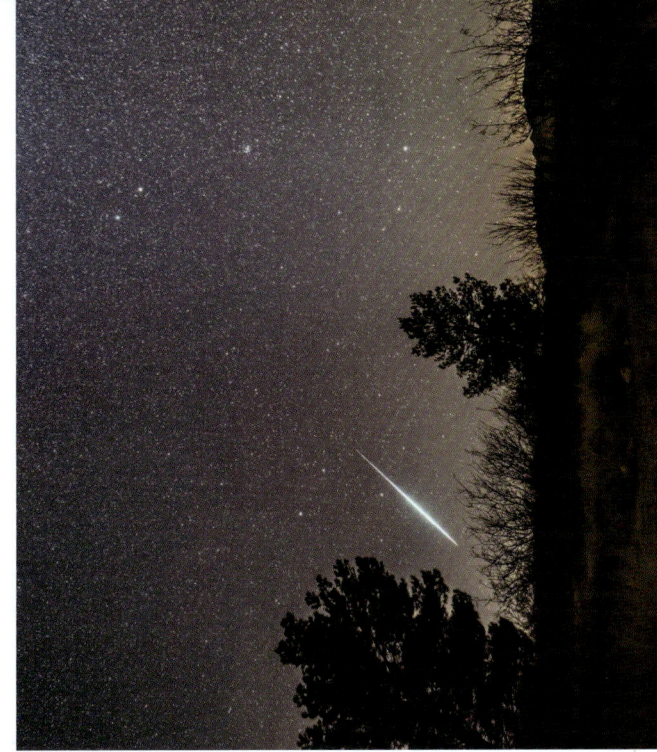

A Geminid fireball. The meteors can travel through the atmosphere at speeds of up to 79,000 miles per hour (127,000 kilometres per hour).

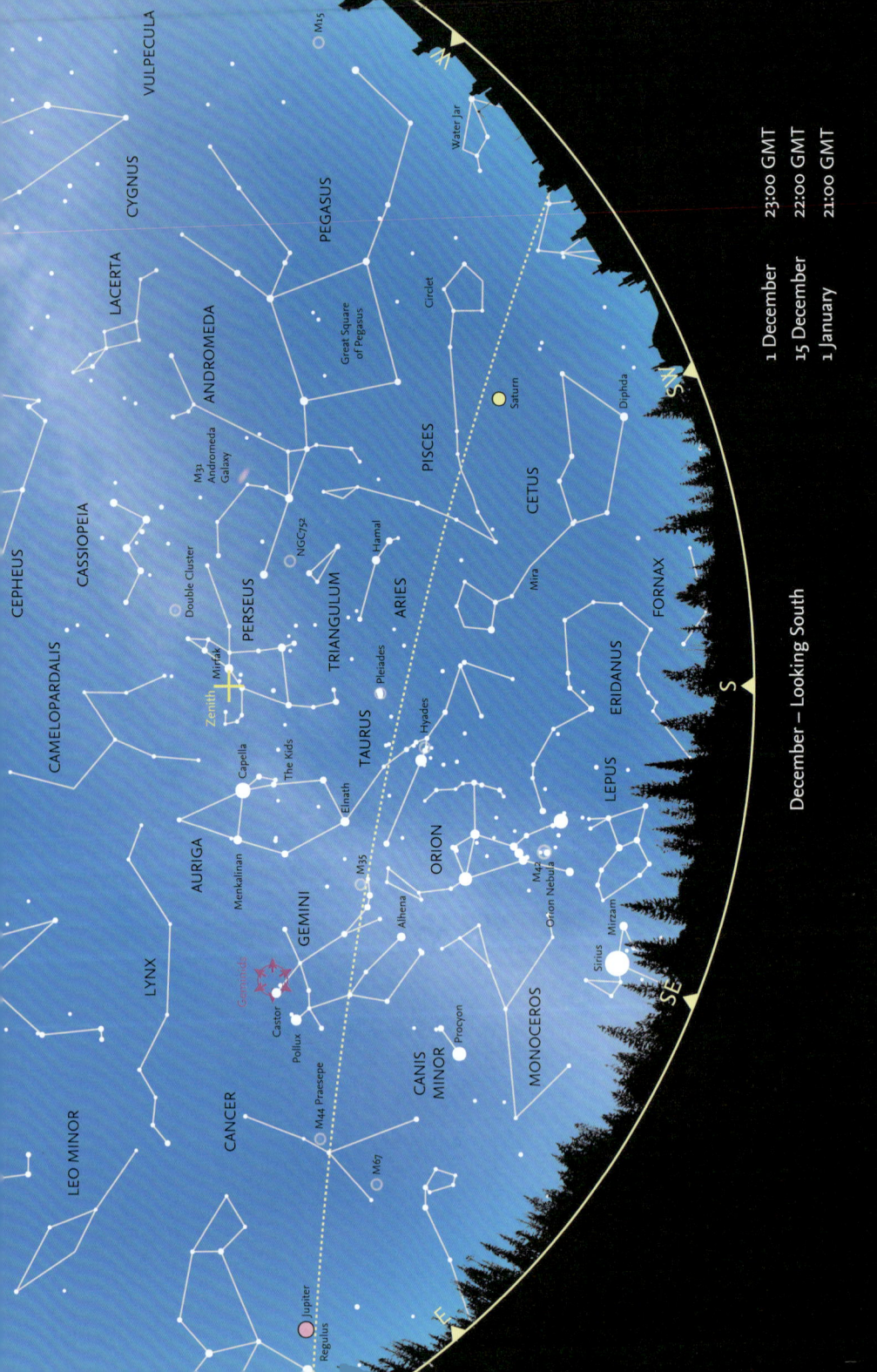

December – Looking South

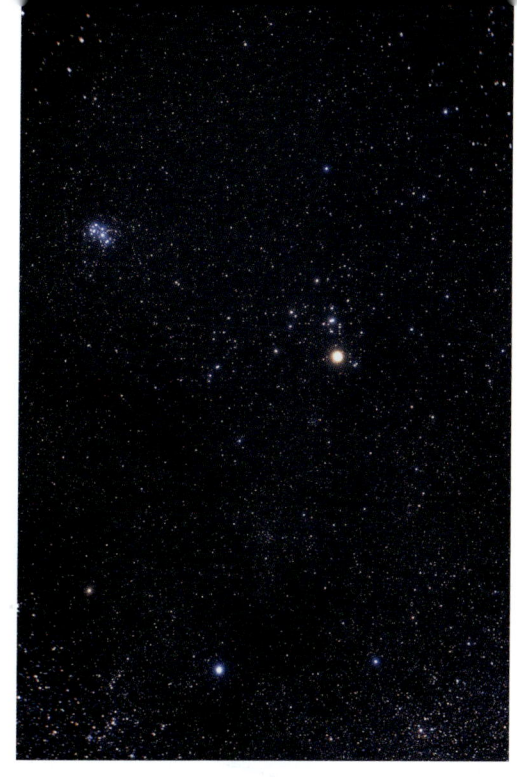

The fine open cluster of the **Pleiades** is due south around 22:00, with the **Hyades** cluster, **Aldebaran** and the rest of **Taurus** clearly visible to the east. **Auriga** (with **Capella**) and **Gemini** (with **Castor** and **Pollux**) are both well placed for observation. **Orion** has made a welcome return to the winter sky, and both **Canis Minor** (with **Procyon**) and **Canis Major** (with **Sirius**, the brightest star in the sky) are now well above the horizon. The small, poorly known constellation of **Lepus** lies to the south of Orion. In the west, **Aquarius** has now disappeared, and **Cetus** is becoming lower, but **Pisces** is still easily seen, as are the constellations of **Aries**, **Triangulum** and **Andromeda** above it. The Great Square of **Pegasus** is starting to plunge down towards the western horizon and, because of its orientation on the sky, appears more like a large diamond standing on one point than a square.

The constellation of Taurus contains two contrasting open clusters: the compact Pleiades, with its striking blue-white stars, and the more scattered, 'V'-shaped Hyades, which are much closer to us. Orange Aldebaran (α Tauri) forms part of the 'V'.

The Moon's phases for December 2026

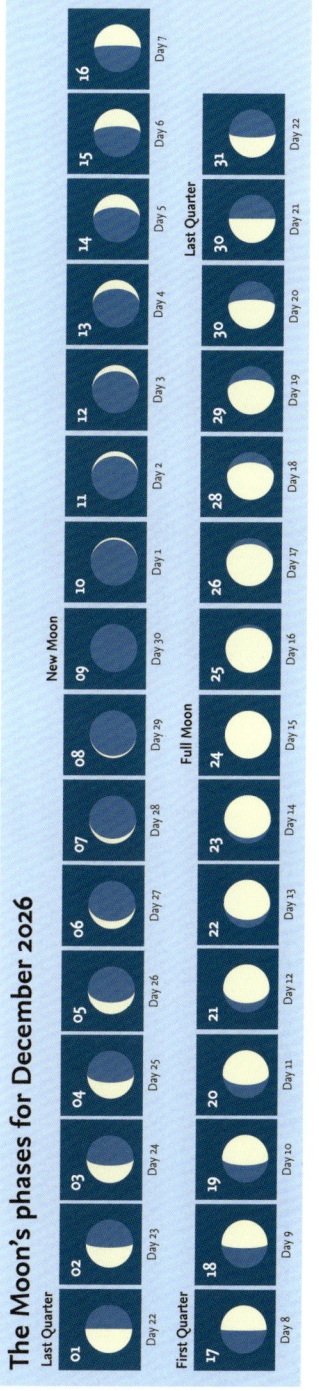

December – Moon and Planets

The Moon

On 1 December the Moon is Last Quarter; as it wanes it moves to 2.5°S of **Spica**. A New Moon arises on 9 December, and 2 days later it is at its farthest point from the Earth (apogee), at a distance of 406,421 km. First Quarter is on 17 December. The waxing gibbous Moon is 1.0°N of the **Pleiades**, on 21 December; three days later it is Full. On 25 December the waning gibbous Moon is 4.4°S of **Pollux**; two days later it moves to 1.5°S of **Jupiter** (mag. -2.4) and 1.4°S of **Regulus** in **Leo**. On 24 December the Moon makes its closest approach to the Earth, with a perigee distance of 356,650 km. The Last Quarter Moon occurs on 30 December.

The Planets

Mercury sits in **Libra** then passes through **Scorpius** and **Ophiuchus** and **Sagittarius**; it approaches the Sun, switching from appearing at dawn to dusk after New Year's Day (mag. -0.7 to -1.3). **Venus** is in **Virgo** visible at dawn, it passes into Libra and reaches western elongation in the new year on 3 January 2027 (mag. -4.9 to -4.6). **Mars** stays in **Leo** for the month; it is above the horizon a few hours before midnight by New Year (mag. 0.4 to -0.1). **Jupiter** is in Leo and visible later in the night with Mars (mag. -2.2 to -2.4). On 12 December Jupiter (mag. -2.3) is 1.3°N of **Regulus**. It begins to move westwards in Leo on 13 December. **Saturn** is on the border between **Cetus** and **Pisces** and visible from sunset to just before midnight (mag. 0.6 to 0.7). On 11 December Saturn ends retrograde motion at mag. 0.6. **Uranus** is settled in **Taurus** and visible all night (mag. 5.6). **Neptune** continues in Pisces at mag. 7.7 to 7.8; it ends retrograde motion on 12 December, moving eastwards towards the end of the year.

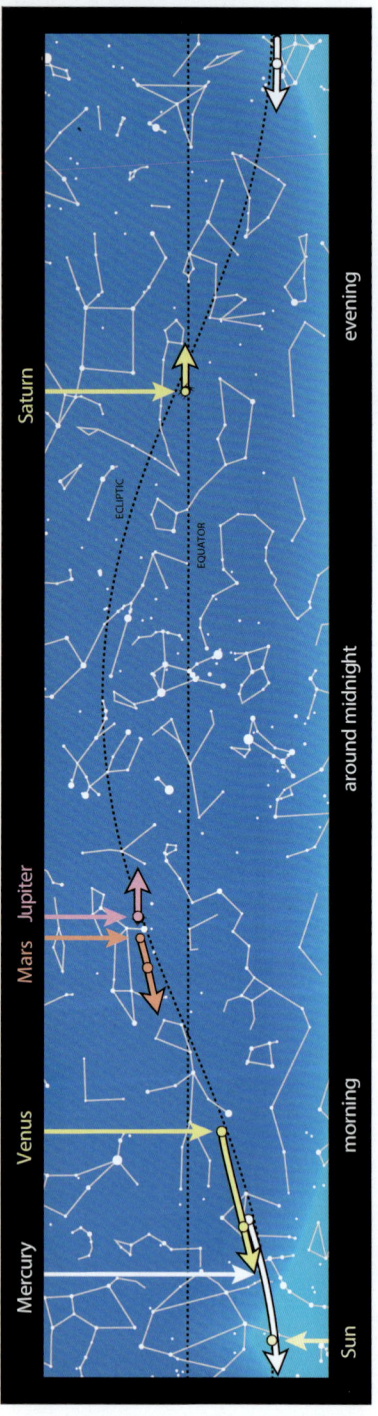

The path of the Sun and the planets along the ecliptic in December.

Calendar for December

01	06:09	Last Quarter Moon
04	18:36	Spica 2.5°N of the Moon
09	00:52	New Moon
11	06:46	Moon at apogee = 406,421 km
12	15:35	Jupiter (mag. -2.3) 1.3°N of Regulus
14		Geminid meteor shower maximum
17	05:43	First Quarter Moon
21	20:50	Winter Solstice
21	22:37	Pleiades 1.0°S of the Moon
22		Ursid meteor shower maximum
24	01:28	Full Moon
24	08:30	Moon at perigee = 356,650 km
25	11:41	Pollux 4.4°N of the Moon
27	17:32	Jupiter (mag. -2.4) 1.5°N of the Moon
27	22:44	Regulus 1.4°N of the Moon
30	18:59	Last Quarter Moon

5 December • Waning crescent Moon close to Venus and Spica, Arcturus visible in the east.

13 December • Jupiter and Regulus appear together, Mars due east in Leo.

25 December • The waning gibbous Moon close to Pollux and Castor, Jupiter and Mars trail in the southwest.

27 December • The waning gibbous Moon next to Regulus, Jupiter and Mars are close by.

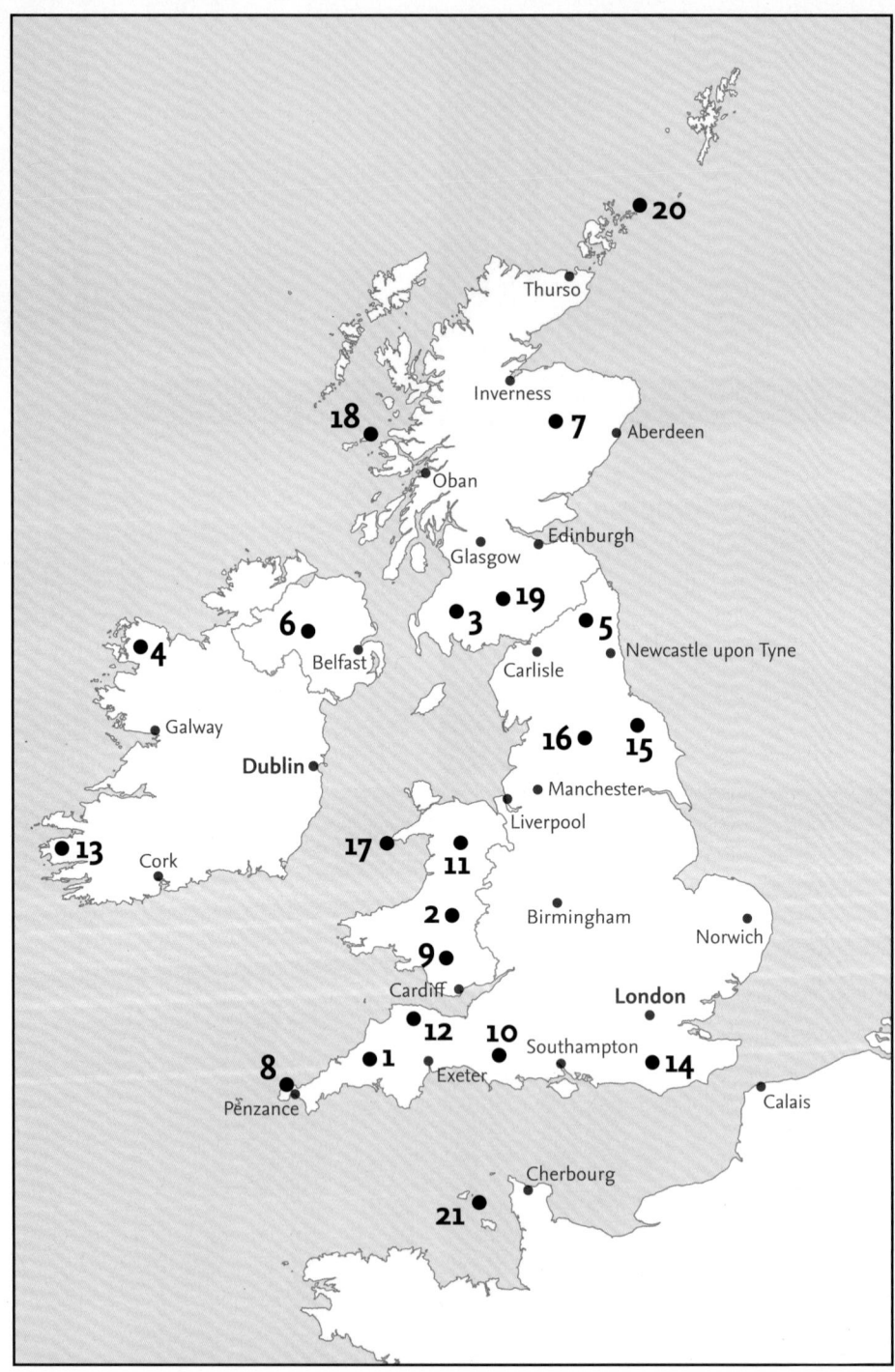

Dark Sky Sites

International Dark-Sky Association Sites

The *International Dark-Sky Association* (IDA) recognizes various categories of sites that offer areas where the sky is dark at night, free from light pollution and particularly suitable for astronomical observing. A number of sites in Great Britain and Ireland have been given specific recognition and are shown on the map. These are:

Parks
1. *Bodmin Moor Dark Sky Landscape*
2. *Elan Valley Estate*
3. *Galloway Forest Park*
4. *Mayo Dark Sky Park*
5. *Northumberland National Park and Kielder Water & Forest Park*
6. *OM Dark Sky Park & Observatory*
7. *Tomintoul and Glenlivet – Cairngorms*
8. *West Penwith*

Reserves
9. *Brecon Beacons National Park*
10. *Cranborne Chase*
11. *Eryri National Park (Snowdonia)*
12. *Exmoor National Park*
13. *Kerry*
14. *Moore's Reserve South Down National Park*
15. *North York Moors National Park*
16. *Yorkshire Dales National Park*

Sanctuaries
17. *Ynys Enlli (Bardsey Island)*

Communities
18. *The Island of Coll (Inner Hebrides)*
19. *Moffat*
20. *North Ronaldsay Dark Sky Island (Orkney Islands)*
21. *The island of Sark (Channel Islands)*

Details of these sites and web links may be found at the IDA website: https://www.darksky.org/ Many of these sites have major observatories or other facilities available for public observing (often at specific dates or times).

Dark Sky Discovery Sites

In Britain there is also the *Dark Sky Discovery* organisation. This gives recognition to smaller sites, again free from immediate light pollution, that are open to observing at any time. Some sites are used for specific, public observing sessions. A full listing of sites is at **https://www.darkskydiscovery.org.uk/** but specific events are publicized locally.

Glossary and Tables

aphelion	The point on an orbit that is farthest from the Sun.
apogee	The point on its orbit at which the Moon is farthest from the Earth.
appulse	The apparently close approach of two celestial objects; two planets, or a planet and star.
astronomical unit	(AU) The mean distance of the Earth from the Sun, 149,597,870 km.
celestial equator	The great circle on the celestial sphere that is in the same plane as the Earth's equator.
celestial sphere	The apparent sphere surrounding the Earth on which all celestial bodies (stars, planets, etc.) seem to be located.
conjunction	The point in time when two celestial objects have the same celestial longitude. In the case of the Sun and a planet, superior conjunction occurs when the planet lies on the far side of the Sun (as seen from Earth). For Mercury and Venus, inferior conjunction occurs when they pass between the Sun and the Earth.
direct motion	Motion from west to east on the sky.
ecliptic	The apparent path of the Sun across the sky throughout the year. Also: the plane of the Earth's orbit in space.
elongation	The point at which an inferior planet has the greatest angular distance from the Sun, as seen from Earth.
equinox	The two points during the year when night and day have equal duration. Also: the points on the sky at which the ecliptic intersects the celestial equator. The vernal (spring) equinox is of particular importance in astronomy.
gibbous	The stage in the sequence of phases at which the illumination of a body lies between half and full. In the case of the Moon, the term is applied to phases between First Quarter and Full, and between Full and Last Quarter.
inferior planet	Either of the planets Mercury or Venus, which have orbits inside that of the Earth.
magnitude	The brightness of a star, planet or other celestial body. It is a logarithmic scale, where larger numbers indicate fainter brightness. A difference of 5 in magnitude indicates a difference of 100 in actual brightness, thus a first-magnitude star is 100 times as bright as one of sixth magnitude.
meridian	The great circle passing through the North and South Poles of a body and the observer's position; or the corresponding great circle on the celestial sphere that passes through the North and South Celestial Poles and also through the observer's zenith.
nadir	The point on the celestial sphere directly beneath the observer's feet, opposite the zenith.
occultation	The disappearance of one celestial body behind another, such as when stars or planets are hidden behind the Moon.
opposition	The point on a superior planet's orbit at which it is directly opposite the Sun in the sky.
perigee	The point on its orbit at which the Moon is closest to the Earth.
perihelion	The point on an orbit that is closest to the Sun.
retrograde motion	Motion from east to west on the sky.
superior planet	A planet that has an orbit outside that of the Earth.
vernal equinox	The point at which the Sun, in its apparent motion along the ecliptic, crosses the celestial equator from south to north. Also known as the First Point of Aries.
zenith	The point directly above the observer's head.
zodiac	A band, streching 8° on either side of the ecliptic, within which the Moon and planets appear to move. It consists of twelve equal areas, originally named after the constellation that once lay within it.

The Constellations

There are 88 constellations covering the whole of the celestial sphere, but 24 of these in the southern hemisphere can never be seen (even in part) from the latitude of Britain and Ireland, so are omitted from this table. The names themselves are expressed in Latin, and the names of stars are frequently given by Greek letters followed by the genitive of the constellation name. The genitives and English names of the various constellations are included.

Name	Genitive	Abbr.	English name	Name	Genitive	Abbr.	English name
Andromeda	Andromedae	And	Andromeda	Lacerta	Lacertae	Lac	Lizard
Antlia	Antliae	Ant	Air Pump	Leo	Leonis	Leo	Lion
Aquarius	Aquarii	Aqr	Water Bearer	Leo Minor	Leonis Minoris	LMi	Little Lion
Aquila	Aquilae	Aql	Eagle	Lepus	Leporis	Lep	Hare
Aries	Arietis	Ari	Ram	Libra	Librae	Lib	Scales
Auriga	Aurigae	Aur	Charioteer	Lupus	Lupi	Lup	Wolf
Boötes	Boötis	Boo	Herdsman	Lynx	Lyncis	Lyn	Lynx
Camelopardalis	Camelopardalis	Cam	Giraffe	Lyra	Lyrae	Lyr	Lyre
Cancer	Cancri	Cnc	Crab	Microscopium	Microscopii	Mic	Microscope
Canes Venatici	Canum Venaticorum	CVn	Hunting Dogs	Monoceros	Monocerotis	Mon	Unicorn
Canis Major	Canis Majoris	CMa	Big Dog	Ophiuchus	Ophiuchi	Oph	Serpent Bearer
Canis Minor	Canis Minoris	CMi	Little Dog	Orion	Orionis	Ori	Orion
Capricornus	Capricorni	Cap	Sea Goat	Pegasus	Pegasi	Peg	Pegasus
Cassiopeia	Cassiopeiae	Cas	Cassiopeia	Perseus	Persei	Per	Perseus
Centaurus	Centauri	Cen	Centaur	Pisces	Piscium	Psc	Fishes
Cepheus	Cephei	Cep	Cepheus	Piscis Austrinus	Piscis Austrini	PsA	Southern Fish
Cetus	Ceti	Cet	Whale	Puppis	Puppis	Pup	Stern
Columba	Columbae	Col	Dove	Pyxis	Pyxidis	Pyx	Compass
Coma Berenices	Comae Berenices	Com	Berenice's Hair	Sagitta	Sagittae	Sge	Arrow
Corona Australis	Coronae Australis	CrA	Southern Crown	Sagittarius	Sagittarii	Sgr	Archer
Corona Borealis	Coronae Borealis	CrB	Northern Crown	Scorpius	Scorpii	Sco	Scorpion
Corvus	Corvi	Crv	Crow	Sculptor	Sculptoris	Scl	Sculptor
Crater	Crateris	Crt	Cup	Scutum	Scuti	Sct	Shield
Cygnus	Cygni	Cyg	Swan	Serpens	Serpentis	Ser	Serpent
Delphinus	Delphini	Del	Dolphin	Sextans	Sextantis	Sex	Sextant
Draco	Draconis	Dra	Dragon	Taurus	Tauri	Tau	Bull
Equuleus	Equulei	Equ	Little Horse	Triangulum	Trianguli	Tri	Triangle
Eridanus	Eridani	Eri	River Eridanus	Ursa Major	Ursae Majoris	UMa	Great Bear
Fornax	Fornacis	For	Furnace	Ursa Minor	Ursae Minoris	UMi	Lesser Bear
Gemini	Geminorum	Gem	Twins	Vela	Velorum	Vel	Sails
Hercules	Herculis	Her	Hercules	Virgo	Virginis	Vir	Virgin
Hydra	Hydrae	Hya	Water Snake	Vulpecula	Vulpeculae	Vul	Fox

The Greek Alphabet

α	Alpha	ε	Epsilon	ι	Iota	ν	Nu	ρ	Rho	φ (φ) Phi
β	Beta	ζ	Zeta	κ	Kappa	ξ	Xi	σ (ς)	Sigma	χ Chi
γ	Gamma	η	Eta	λ	Lambda	ο	Omicron	τ	Tau	ψ Psi
δ	Delta	θ (ϑ)	Theta	μ	Mu	π	Pi	υ	Upsilon	ω Omega

GLOSSARY AND TABLES 109

Some common asterisms

Belt of Orion	δ, ε and ζ Orionis
Big Dipper	α, β, γ, δ, ε, ζ and η Ursae Majoris
Circlet	γ, θ, ι, λ and κ Piscium
Guards (or Guardians)	β and γ Ursae Minoris
Head of Cetus	α, γ, ξ², μ and λ Ceti
Head of Draco	β, γ, ξ and ν Draconis
Head of Hydra	δ, ε, ζ, η, ρ and σ Hydrae
Keystone	ε, ζ, η and π Herculis
Kids	ζ and η Aurigae
Little Dipper	β, γ, η, ζ, ε, δ and α Ursae Minoris
Lozenge	= Head of Draco
Milk Dipper	ζ, γ, σ, φ and λ Sagittarii
Plough or Big Dipper	α, β, γ, δ, ε, ζ and η Ursae Majoris
Pointers	α and β Ursae Majoris
Sickle	α, η, γ, ζ, μ and ε Leonis
Square of Pegasus	α, β and γ Pegasi with α Andromedae
Sword of Orion	θ and ι Orionis
Teapot	γ, ε, δ, λ, φ, σ, τ and ζ Sagittarii
Wain (or Charles' Wain)	= Plough
Water Jar	γ, η, κ and ζ Aquarii
Y of Aquarius	= Water Jar

Acknowledgements

Our thanks to Ed Bloomer, Imo Bell, Catherine Muller and Sam Imperato, astronomers at Royal Observatory Greenwich.

Image Credits:

pp. 16, 21, 32, 35, 73, 95, 97 Shutterstock
p. 28 Dr Russell Cockman, www.russellsastronomy.com
p. 29 Christian Gloor, CC BY 2.0
p. 37 The Starmon, CC BY 3.0
p. 41 pithecanthropus4152, CC BY-SA 4.0
p. 43 Stephen Rahn from Macon, GA, USA, CCo
p. 47 NiKo, CC BY-SA 4.0
p. 49 Fried Lauterbach, CC BY-SA 4.0
p. 53 PsamatheM, CC BY-SA 4.0
p. 55 Roberto Mura, CC BY-SA 3.0
p. 59 Genuson, CC BY-SA 3.0
p. 61 Roberto Mura, CC BY-SA 3.0
p. 65 Brwynog, CC BY-SA 4.0
p. 67 Album/Alamy Stock Photo
p. 71 David Ritter, CC BY-SA 4.0
p. 77 mLu.fotos, CC BY 2.0
p. 79 Starhopper, CC BY-SA 4.0
p. 83 Keesscherer, CC BY-SA 4.0
p. 85 World History Archive/Alamy Stock Photo
p. 101 Stocktrek Images Inc/Alamy Stock Photo
p. 103 Akira Fujii/ESA/Hubble

Further Information

Books
Bone, Neil (1993), *Observer's Handbook: Meteors*, George Philip, London & Sky Publ. Corp., Cambridge, Mass.
Chu, A (2012), *The Cambridge Photographic Moon Atlas*, Cambridge University Press, Cambridge
Dunlop, Storm (1999), *Wild Guide to the Night Sky*, HarperCollins, London
Dunlop, Storm (2012), *Practical Astronomy*, 3rd edn, Philip's, London
Dunlop, Storm, Rükl, Antonin & Tirion, Wil (2005), *Collins Atlas of the Night Sky*, HarperCollins, London
Ellyard, David & Tirion, Wil (2008), *Southern Sky Guide*, 3rd edn, Cambridge University Press, Cambridge
Heifetz, Milton & Tirion, Wil (2017), *A Walk through the Heavens*, 4th edn, Cambridge University Press, Cambridge
Heifetz, Milton & Tirion, Wil (2012), *A Walk through the Southern Sky*, 3rd edn, Cambridge University Press, Cambridge
National Geographic & Wei-Haas, Maya (2023), *Stargazer's Atlas: The Ultimate Guide to the Night Sky*, National Geographic, Washington, D.C.
O'Meara, Stephen J. (2008), *Observing the Night Sky with Binoculars*, Cambridge University Press, Cambridge
Ridpath, Ian (2018), *Star Tales*, 2nd edn, Lutterworth Press, Cambridge
Ridpath, Ian, ed. (2003), *Oxford Dictionary of Astronomy*, 2nd edn, Oxford University Press, Oxford
Ridpath, Ian, ed. (2004), *Norton's Star Atlas*, 20th edn, Pi Press, New York
Ridpath, Ian & Tirion, Wil (2004), *Collins Gem – Stars*, HarperCollins, London
Ridpath, Ian & Tirion, Wil (2017), *Collins Pocket Guide Stars and Planets*, 5th edn, HarperCollins, London
Ridpath, Ian & Tirion, Wil (2019), *The Monthly Sky Guide*, 10th edn, Dover Publications, New York
Rükl, Antonín (1990), *Hamlyn Atlas of the Moon*, Hamlyn, London & Astro Media Inc., Milwaukee
Rükl, Antonín (2004), *Atlas of the Moon*, Sky Publishing Corp., Cambridge, Mass.
Scagell, Robin (2000), *Philip's Stargazing with a Telescope*, George Philip, London
Sky & Telescope (2017), *Astronomy 2018*, Australian Sky & Telescope, Quasar Publishing, Georges Hall, NSW
Stimac, Valerie (2019) *Dark Skies: A Practical Guide to Astrotourism*, Lonely Planet, Franklin, TN
Tirion, Wil (2011), *Cambridge Star Atlas*, 4th edn, Cambridge University Press, Cambridge
Topalovic, Radmila & Kerss, Tom (2016), *Stargazing: Beginner's Guide to Astronomy*, HarperCollins, London

Journals
Astronomy, Astro Media Corp., 21027 Crossroads Circle, P.O. Box 1612, Waukesha, WI 53187-1612 USA. http://www.astronomy.com
Astronomy Now, Pole Star Publications, PO Box 175, Tonbridge, Kent TN10 4QX UK. http://astronomynow.com
Sky at Night Magazine, BBC publications, London. http://skyatnightmagazine.com
Sky & Telescope, Sky Publishing Corp., Cambridge, MA 02138-1200 USA. http://www.skyandtelescope.org/

Societies
British Astronomical Association, Burlington House, Piccadilly, London W1J 0DU. http://www.britastro.org/
The principal British organization for amateur astronomers (with some professional members), particularly for those interested in carrying out observational programmes. Its membership is, however, worldwide. It publishes fully refereed, scientific papers and other material in its well-regarded journal.

Federation of Astronomical Societies, Secretary: Ken Sheldon, Whitehaven, Maytree Road, Lower Moor, Pershore, Worcs. WR10 2NY. http://www.fedastro.org.uk/fas/
An organization that is able to provide contact information for local astronomical societies in the United Kingdom.

Royal Astronomical Society, Burlington House, Piccadilly, London W1J 0BQ. http://www.ras.org.uk/
The premier astronomical society, with membership primarily drawn from professionals and experienced amateurs. It has an exceptional library and is a designated centre for the retention of certain classes of astronomical data. Its publications are the standard medium for dissemination of astronomical research.

Society for Popular Astronomy, 36 Fairway, Keyworth, Nottingham NG12 5DU.
http://www.popastro.com/
A society for astronomical beginners of all ages, which concentrates on increasing members' understanding and enjoyment, but which does have some observational programmes. Its journal is entitled *Popular Astronomy*.

Software

Planetary, Stellar and Lunar Visibility (planetary and eclipse freeware): Alcyone Software, Germany. http://www.alcyone.de
Redshift, Redshift-Live. http://www.redshift-live.com/en/
Starry Night & Starry Night Pro, Sienna Software Inc., Toronto, Canada. http://www.starrynight.com
Stellarium, https://stellarium.org/

Internet sources

There are numerous sites with information about all aspects of astronomy, and all of those have numerous links. Although many amateur sites are excellent, treat any statements and data with caution. The sites listed below offer accurate information. Please note that the URLs may change. If so, use a good search engine, such as Google, to locate the information source.

Information

Astronomical data (inc. eclipses) HM Nautical Almanac Office: http://astro.ukho.gov.uk
Auroral information Michigan Tech: http://www.geo.mtu.edu/weather/aurora/
Comets JPL Solar System Dynamics: http://ssd.jpl.nasa.gov/
American Meteor Society: http://amsmeteors.org/
Deep-sky objects Saguaro Astronomy Club Database: http://www.virtualcolony.com/sac/
Eclipses NASA Eclipse Page: http://eclipse.gsfc.nasa.gov/eclipse.html
Ice in Space (Southern Hemisphere Online Astronomy Forum): http://www.iceinspace.com.au/index.php?home
Moon (inc. Atlas) Inconstant Moon: http://www.inconstantmoon.com/
Planets Planetary Fact Sheets: http://nssdc.gsfc.nasa.gov/planetary/planetfact.html
Satellites (inc. International Space Station)
    Heavens Above: http://www.heavens-above.com/
    Visual Satellite Observer: http://www.satobs.org/
Star Chart http://www.skyandtelescope.com/observing/interactive-sky-watching-tools/interactive-sky-chart/
What's Visible
    Skyhound: http://www.skyhound.com/sh/skyhound.html
    Skyview Cafe: http://www.skyviewcafe.com

Institutes and Organizations

European Space Agency: http://www.esa.int/
International Dark-Sky Association: http://www.darksky.org/
RASC Dark Sky: https://rasc.ca/
Jet Propulsion Laboratory: http://www.jpl.nasa.gov/
Lunar and Planetary Institute: http://www.lpi.usra.edu/
National Aeronautics and Space Administration: http://www.hq.nasa.gov/
Solar Data Analysis Center: http://umbra.gsfc.nasa.gov/
Space Telescope Science Institute: http://www.stsci.edu/